CLIMATE CHANGE AND ITS CAUSES, EFFECTS AND PREDICTION

CLIMATE CHANGE

BACKGROUND, FUNDING AND IMPACTS

Climate Change and its Causes, Effects and Prediction

Additional books and e-books in this series can be found on Nova's website under the Series tab.

CLIMATE CHANGE AND ITS CAUSES, EFFECTS AND PREDICTION

CLIMATE CHANGE

BACKGROUND, FUNDING AND IMPACTS

JEREMY NESBITT
EDITOR

New York

Library of Congress Cataloging-in-Publication Data

ISBN: 978-1-53617-216-4

Published by Nova Science Publishers, Inc. † New York

CONTENTS

PREFACE

Regarding climate changes, a key question has been the degree to which humans and natural factors have influenced observed global climate change. Chapter 1 traces the evolution of scientific understanding and confidence regarding the drivers of recent global climate change. Chapter 2 reviews how U.S. agencies address climate change as a potential driver of global migration. Chapter 3 examines reported federal funding from 2010 to 2017 and the extent to which reports on such funding are clearly linked to the federal fiscal exposure to climate change; the extent to which selected agencies reported climate change funding that supports programs where addressing climate change is the primary purpose; and the extent to which the primary purpose programs are fragmented, overlapping, or duplicative.

The costs of recent weather disasters have illustrated the need for planning for climate change risks and investing in resilience as reported in chapter 4.

For more than a decade, federal agencies have grappled with how to address climate change effects when implementing the Endangered Species Act of 1973 (ESA). As set forth by Congress, one of the main purposes of the ESA is to "provide a means whereby the ecosystems upon which endangered species and threatened species depend may be conserved." Chapter 5 analyzes the courts' role in shaping how the Services have factored climate change effects into ESA decisions and recent 2019 regulatory developments that aim to clarify how the Services consider and address climate change in their ESA decisions.

Many governments hold that environmental degradation and climate change pose international and trans-boundary risks to human populations, economies, and ecosystems. International financial assistance, or foreign aid, has been a principal method for governments to support actions on global environmental problems in lower-income countries. As discussed in chapter 6 and 7, this assistance may include grants, loans, loan guarantees, export credits, insurance products, and private sector investment.

On June 1, 2017, President Trump announced his intent to withdraw the United States from the Paris Agreement (PA), an international accord to address climate change over the coming century. Some observers argue that the Administration's decision to withdraw from the PA will (1) reduce the U.S. standing in the world by making the United States an international outlier on climate change, (2) strengthen perceptions that the United States is withdrawing from its traditional position of world leadership and becoming more inward-focused or even isolationist, (3) create an opportunity for China to assume a position of world leadership on climate change and perhaps other issues, and (4) make the United States appear less reliable as a negotiating partner, which could make it harder for the United States in the future to secure foreign cooperation for addressing other issues of mutual interest or to call on other countries to abide by their commitments in other international agreements as reported in chapter 8.

Surface transportation is a major source of carbon dioxide (CO2) in the atmosphere, the main human-related greenhouse gas (GHG) contributing to climate change. At the same time, the effects of climate change, such as extreme heat, sea level rise, and stronger storms, pose a threat to transportation infrastructure. Chapter 9 seeks to address these two aspects of climate change with mitigation provisions that aim to reduce GHG emissions from surface transportation and adaptation provisions that aim to make the surface transportation system more resilient to a changing climate.

For policymakers considering actions to reduce GHG emissions, various policy instruments are available. Over the last 15 years, many legislative proposals have involved market-based approaches, such as a GHG emissions cap-and-trade system or a carbon tax. These particular approaches may be considered in the 116th Congress and are discussed in chapter 10.

As mandated by Congress, the U.S. Global Change Research Program produces regular National Climate Assessment (NCA) reports on the state of scientific knowledge about climate change and its effects on human and natural systems in the United States. According to NCA4, "annual losses in some economic sectors are projected to reach hundreds of billions of dollars by the end of the century" provided continued growth in emissions at historic rates. The findings of the NCA4 are the focus of chapter 11.

In light of public concern over climate change, some stakeholders have asked to what extent publicly traded companies should disclose their climate-related risks. As reported in chapter 12, while current SEC requirements do not address climate-related risks expressly, publicly traded companies must disclose such risks if they are "material" under federal securities laws.

Chapter 1 - This CRS report provides context for the Administration's Climate Science Special Report (October 2017) by tracing the evolution of scientific understanding and confidence regarding the drivers of recent global climate change.

Chapter 2 - The effects of climate change, combined with other factors, may alter human migration trends across the globe, according to the International Organization for Migration. For example, climate change can increase the frequency and intensity of natural disasters, causing populations to move from an area. Climate change can also intensify slow-onset disasters, such as drought, crop failure, or sea level rise, potentially altering longer-term migration trends. GAO was asked to review how U.S. agencies address climate change as a potential driver of global migration. For State, USAID, and DOD, this chapter (1) describes executive branch actions related to climate change and migration from fiscal years 2014 through 2018; (2) examines the extent to which the agencies discussed the potential effects of climate change on migration in their plans and risk assessments; and (3) describes agency activities on the issue. GAO analyzed documents on administration priorities; reviewed agency plans, risk assessments, and documentation of agency activities; and interviewed agency officials.

Chapter 3 - Since 1993, OMB has reported over $154 billion in funding for federal climate change activities, spread across the government—raising questions about fragmentation, overlap, or duplication. GAO was asked to review federal climate change funding. This chapter examines (1) reported federal funding from 2010 to 2017 and the extent to which reports on such funding are clearly linked to the federal fiscal exposure to climate change; (2) the extent to which selected agencies reported climate change funding that supports programs where addressing climate change is the primary purpose; and (3) the extent to which the primary purpose programs are fragmented, overlapping, or duplicative. GAO reviewed OMB climate change funding reports; analyzed budget justifications for six agencies—the Departments of Agriculture, Commerce, Defense, and Energy; the National Aeronautics and Space Administration and the National Science Foundation—representing 89 percent of OMB-reported climate change funding in fiscal year 2014; analyzed documents on primary purpose programs against GAO's fragmentation, overlap, or duplication criteria; and reviewed GAO's prior work on fiscal exposures.

Chapter 4 - Since 2005, federal funding for disaster assistance is at least $450 billion, including approximately $19.1 billion in supplemental appropriations signed into law on June 6, 2019. In 2018 alone, there were 14 separate billion-dollar weather and climate disaster events across the United States, with a total cost of at least $91 billion, according to the National Oceanic and Atmospheric Administration. The U.S. Global Change Research Program projects that disaster costs will likely increase as certain extreme weather events become more frequent and intense due to climate change. The costs of recent weather disasters have illustrated the need for planning for climate change risks and investing in resilience. Resilience is the ability to prepare and plan for, absorb, recover from, and more successfully adapt to adverse events, according to the National Academies of Science, Engineering, and Medicine. Investing in resilience can reduce the need for far more costly steps in the decades to come. Since February 2013, GAO has included *Limiting the Federal Government's Fiscal Exposure by Better Managing Climate Change Risks* on its list of federal program areas at high risk of vulnerabilities to fraud, waste, abuse, and mismanagement or most in need of transformation. GAO updates this list every 2 years. In March 2019, GAO reported that the federal government had not made measurable progress since 2017 to reduce fiscal exposure to climate change. This testimony—based on reports GAO issued from October 2009 to March 2019—discusses (1) what is known about the potential economic effects of climate change in the United States and the extent to which this information could help federal decision makers manage climate risks across the federal government, (2) the potential impacts of climate change on the federal budget, (3) the extent to which the federal government has invested in resilience, and (4) how the federal government could reduce fiscal exposure to the effects of climate change. GAO has made 62 recommendations related to the *Limiting the Federal Government's Fiscal Exposure by Better Managing Climate Change Risks* high-risk area. As of December 2018, 25 of those recommendations remained open.

Chapter 5 - For more than a decade, federal agencies have grappled with how to address climate change effects when implementing the Endangered Species Act of 1973 (ESA). The ESA aims to protect threatened and endangered fish, wildlife, and plants from extinction. As set forth by Congress, one of the main purposes of the ESA is to "provide a means whereby the ecosystems upon which endangered species and threatened species depend may be conserved." The U.S. Fish and Wildlife Service (FWS) and the National Marine Fisheries Service (NMFS) (collectively, the Services) have acknowledged that the changing climate may threaten the survival of and habitat for some species. As noted by courts and legal scholars, the ESA does not expressly require the Services to consider the effect of climate change in their ESA decisions. However, the ESA and its implementing regulations (1) direct the Services to consider "natural or manmade factors affecting [a species'] continued existence" when determining whether a species should be protected under the ESA; and (2) require the Services to analyze cumulative effects on a species' survival when analyzing whether federal actions jeopardize a species protected under the Act. The courts and the Services have interpreted these provisions as requiring the Services to consider climate change effects in the ESA decisionmaking process. Various lawsuits have challenged the Services' interpretation of complex scientific data or models that predict short- and long-term effects from a changing global climate on specific species and their habitats. Legal challenges have influenced how the Services implement the ESA when climate change affects species and their habitats. Lawsuits typically focus on two main issues: (1) when the Services should list, delist, or reclassify a species as threatened or endangered because of climate change effects; and (2) whether the Services can or should regulate activities that affect the climate to protect species and their habitat. Judicial review has helped to ensure that the Services consider projected climate change effects on species in their ESA decisions. However, the courts have not required the Services to curb activities that may contribute to climate change to protect threatened or endangered species. Stakeholders disagree on whether the ESA should play a role in addressing climate change, with some arguing that the ESA is not equipped to mitigate climate change effects. Other stakeholders believe that the Services can and should wield the ESA to protect further species threatened by climate change by curbing activities contributing to climate change. From the Services' viewpoint, the best available scientific and commercial data have been insufficient to determine whether greenhouse gas emissions from a proposed activity cause detrimental effects on a species or its habitat. In light of the judicial deference afforded to the Services, the courts have not expanded the ESA as a tool to protect listed species by regulating activities that contribute to climate change. This chapter analyzes the courts' role in shaping how the Services have factored climate change effects into ESA decisions and recent 2019 regulatory developments that aim to clarify how the Services consider and address climate change in their ESA decisions. In August 2019, the Services finalized revisions to the ESA implementing regulations, aiming to increase transparency and effectiveness of the ESA while easing regulatory burdens. Among those changes, the Services clarified their existing policies and practices for factoring climate change effects into their ESA decisions. As legislative proposals to revise the ESA continue to develop, legal battles over the how the Services interpret climate change effects in their ESA decisions will likely continue.

Chapter 6 - Many governments hold that environmental degradation and climate change pose international and trans-boundary risks to human populations, economies, and ecosystems. To confront these challenges, governments have negotiated various international agreements to protect the environment, reduce pollution, conserve natural resources, and promote sustainable growth. While some observers call upon industrialized countries to take the lead in addressing these issues, many recognize that efforts are unlikely to be sufficient without similar measures being taken in lower- income countries. However, lower-income countries, which tend to focus on poverty reduction and economic growth, may not have the financial resources, technological know- how, and/or institutional capacity to deploy environmentally protective measures on their own. Therefore, international financial assistance, or foreign aid, has been a principal method for governments to support actions on global environmental problems in lower-income countries. Often, this assistance can serve as a cost-effective strategy for donor countries to provide greater market access for domestic goods and services abroad and increased environmental benefits at home. The United States and other industrialized countries have committed to providing financial assistance for global environmental initiatives through a variety of multilateral agreements, including the Montreal Protocol (1987), the U.N. Framework Convention on Climate Change (1992), and the U.N. Convention to Combat Desertification (1994). International financial assistance takes many forms, from fiscal transfers to market transactions. It may include grants, loans, loan guarantees, export credits, insurance products, and private sector investment. It may be structured as official bilateral development assistance or as contributions to multilateral development banks and other international financial institutions.

Chapter 7 - Many governments hold that environmental degradation and climate change pose international and trans-boundary risks to human populations, economies, and ecosystems. To confront these challenges, governments have negotiated various international agreements to protect the environment, reduce pollution, conserve natural resources, and promote sustainable growth. While some observers call upon industrialized countries to take the lead in addressing these issues, many recognize that efforts are unlikely to be sufficient without similar measures being taken in lower- income countries. However, lower-income countries, which face poverty and economic growth issues, may not have the financial resources, technological know-how, and/or institutional capacity to deploy environmentally protective measures on their own. Therefore, international financial assistance, or foreign aid, has been a principal method for higher-income governments to support actions on global environmental problems in lower-income countries. Often, this assistance can serve as a cost-effective strategy for donor countries to provide greater market access for their environmental goods and services abroad and increased environmental benefits at home. The United States and other industrialized countries have committed to providing financial assistance for global environmental initiatives through a variety of multilateral agreements, including, among others, the Montreal Protocol (1987), the United Nations Framework Convention on Climate Change (UNFCCC, 1992), and the U.N. Convention to Combat Desertification (1994). International financial assistance takes many forms, from fiscal transfers to market transactions. It may include grants, loans, loan guarantees, export credits, insurance products, and private sector investment. It may be structured as official bilateral development assistance or as contributions to multilateral development banks and other international financial institutions. Ultimately, U.S. government assistance to developing countries for environmental initiatives is determined by Congress. Congressional committees of jurisdiction include the House Committees on Foreign Affairs, Financial Services, and Appropriations and the Senate Committees on Foreign Relations and Appropriations.

Chapter 8 - This is an edited, reformatted and augmented version of Congressional Research Service Publication No. IF10668, updated April 5, 2019.

Chapter 9 - Highway, public transportation, and rail programs are authorized through FY2020 by the Fixing America's Surface Transportation (FAST) Act (P.L. 114-94). In August 2019, the Senate Committee on Environment and Public Works reported a bill that would reauthorize the highway elements of surface transportation programs from FY2021 through FY2025. The bill, S. 2302, the America's Transportation Infrastructure Act of 2019, would be the first surface transportation authorization act to include major provisions that address climate change.

Surface transportation is a major source of carbon dioxide (CO2) in the atmosphere, the main human-related greenhouse gas (GHG) contributing to climate change. At the same time, the effects of climate change, such as extreme heat, sea level rise, and stronger storms, pose a threat to transportation infrastructure. S. 2302 seeks to address these two aspects of climate change with mitigation provisions that aim to reduce GHG emissions from surface transportation and adaptation provisions that aim to make the surface transportation system more resilient to a changing climate.

Chapter 10 - Almost all climate scientists agree that greenhouse gas (GHG) emission increases have contributed to observed climate change and that continued increases in GHG emissions will contribute to future climate change.

Although a variety of efforts seeking to reduce GHG emissions are currently underway on the international level and in individual states or regional partnerships, federal policymakers and stakeholders have different viewpoints over what to do—if anything—about GHG emissions, future climate change, and related impacts.

For policymakers considering actions to reduce GHG emissions, various policy instruments are available. Over the last 15 years, many legislative proposals have involved market-based approaches, such as a GHG emissions cap- and-trade system or a carbon tax. These particular approaches may be considered in the 116th Congress and are discussed below. The information below provides an overview of two approaches while briefly addressing their similarities and differences.

Chapter 11 - As mandated by Congress, the U.S. Global Change Research Program produces regular National Climate Assessment (NCA) reports on the state of scientific knowledge about climate change and its effects on human and natural systems in the United States. Volume II of NCA4, published in 2018, examines the contributions of individual economic studies, in order to develop consensus conclusions from across the body of climate impacts literature. According to NCA4, "annual losses in some economic sectors are projected to reach hundreds of billions of dollars by the end of the century" provided continued growth in emissions at historic rates. The findings of the NCA4, and those of previous reports, have raised concerns in Congress about the projected economic impacts and questions about the study methods used to derive these projections.

Studies projecting the economic impacts of climate change vary in their scope, methodology, and the intended applications of their findings. Some focus on the effects of one aspect of climate change upon a single type of economic activity (e.g., effects of temperature and carbon dioxide fertilization on corn yields), while others study the effects across a more comprehensive array of market and non-market activities. Recent research, however, provides detailed projections of future impacts of climate change across a variety of sectors of the U.S. economy.

Projections of the economic impacts of climate change have been cited by some Members of Congress to advance climate-related legislation, and have raised questions and concerns about their interpretation and use in policymaking. This product provides an abbreviated overview of climate impacts research, recent projections from the literature, and considerations in interpreting these. In this context, climate impacts are defined as the projected physical and economic effects of climate change, both positive and negative. This product does not discuss the social cost of carbon or similar metrics, which estimate the net present costs associated with emitting additional greenhouse gases (see CRS In Focus IF10625, Social Costs of Carbon/Greenhouse Gases: Issues for Congress).

Chapter 12 - In light of public concern over climate change, some stakeholders have asked to what extent publicly traded companies should disclose their climate-related risks. The Securities and Exchange Commission (SEC) requires publicly traded companies to disclose financial statements and certain other relevant business information in public filings, including annual and quarterly reports. While current SEC requirements do not address climate-related risks expressly, publicly traded companies must disclose such risks if they are "material" under federal securities laws.

In: Climate Change
Editor: Jeremy Nesbitt

ISBN: 978-1-53617-216-4

Chapter 1

EVOLVING ASSESSMENTS OF HUMAN AND NATURAL CONTRIBUTIONS TO CLIMATE CHANGE*

Jane A. Leggett

SUMMARY

This CRS report provides context for the Administration's Climate Science Special Report (October 2017) by tracing the evolution of scientific understanding and confidence regarding the drivers of recent global climate change.

INTRODUCTION

As Congress continues to deliberate whether and how to address climate change, a key question has been the degree to which humans and

* This is an edited, reformatted and augmented version of a Congressional Research Service publication R45086, prepared for Members and Committees of Congress dated February 1, 2018.

natural factors have influenced observed global climate change.[1] Members of Congress sometimes stress that policies or actions "must be based on sound science."[2] Officials in the Trump Administration have expressed uncertainty about the human influence, and some have called for public debate on the topic.[3]

To help inform policymaking, researchers and major scientific assessment processes have analyzed the attribution of observed climate change to various possible causes. Scientific assessments of both climate change and the extent to which humans have influenced it have varied in expressed confidence over time but have achieved greater scientific consensus. The latest major U.S. assessment, the Climate Science Special Report (CSSR), was released in October 2017 by the U.S. Global Change Research Program (USGCRP). It stated

> It is extremely likely [> 95% likelihood] that human influence has been the dominant cause of the observed warming since the mid-20th Century. For the warming over the last century, there is no convincing

---

[1] Rep. Lamar Smith, in U.S. Congress, House Committee on Science, Space, and Technology, *Climate Science: Assumptions, Policy Implications, and the Scientific Method*, 115th Cong., 1st sess., March 29, 2017, Serial No. 115-10 (Washington: GPO, 2017), p. 4; Sen. John Barrasso, as quoted in "Republican Leader: Climate Change Science 'Not Known,'" *The Hill*, June 4, 2014, http://thehill.com/policy/energy-environment/208431-gop-leader-climate-change- science-not-known; See also the discussion of Senate votes in 2015 regarding climate change attribution to human activities in Dustin Weaver, "Senate Votes That Climate Change Is Real," *The Hill*, January 21, 2015, http://thehill.com/policy/energy-environment/230316-senate-votes-98-1-that-climate-change-is-real.

[2] Ibid. The full quotation from Rep. Smith is as follows:

I believe the climate is changing and that humans play a role. However, I also believe significant questions remain as to the extent. Our actions must be based on sound science. This is the only way we will be able to better address climate change.

[3] For example, the Administrator of the U.S. Environmental Protection Agency, Scott Pruitt, told Fox News, "Human activity contributes to that [climate] change in some measure.... The issue is how much we contribute to it from a human activity perspective and then what can be done about it from a process perspective." Quoted in Anna Giaritelli, "Trump EPA Chief: Humans Contribute to Global Warming, But How Much?," April 2, 2017, http://www.foxnews.com/politics/2017/04/02/trump-epa-chief-humans-contribute-to-global-warming-but-how- much.html. It is presently unclear whether the Administration might proceed with a "Red Team, Blue Team" debate on climate science, as proposed by Administrator Pruitt, continue with the periodic science assessments required by the Global Change Research Act of 1990 (P.L. 110-606), or pursue another approach. See, for example, Robin Bravender, "Climate Science Debate 'On Hold' After White House Meeting," ClimateWire, December 15, 2017, http://thehill.com/policy/energy-environment/365104-trump-admin-puts-epa-climate-science-debate-plan-on-hold.

alternative explanation supported by the extent of the observational evidence.[4]

This chapter provides context for the CSSR's statement by tracing the evolution of scientific understanding and confidence regarding the drivers of recent global climate change.[5]

Climate change science can be traced back to the early 1800s. Through the 20th century, academic institutions, federal and state agencies, foreign governments, and other entities invested significant time and billions of dollars in climate research. This investment has led to substantial advances in empirical observations, atmospheric and ocean physics and chemistry, climate and economic simulation models, statistical methods, and other achievements. As a result, scientists have increased their confidence in their detection and understanding of climate change and attribution of observed changes to their causes. There is now high scientific confidence that the global climate is warming, primarily as a result of increased human-related greenhouse gas (GHG) emissions and other activities.[6] This

---

[4] Donald J. Weubbles et al., *Climate Science Special Report: Fourth National Climate Assessment, Volume I*, USGCRP, 2017, https://science2017.globalchange.gov/ downloads/ CSSR2017_FullReport.pdf. A brief explanation of development of the CSSR is available at https://www.globalchange.gov/content/cssr. The CSSR is the first volume of the fourth national assessment required periodically by the Global Change Research Act of 1990 (P.L. 101-606).

[5] Scientists and commentators often discuss climate change referring to different periods of time. Such time differences can explain some or all of the differences among some perspectives. The factors having greatest influence on global climate typically vary depending on whether the period of interest is one decade, several decades to a century, or tens of thousands to millions of years. While scientific assessments of the drivers of climate change cover all time scales, most assessments consider the human contribution since the Industrial Revolution beginning in the 19th century.

[6] Weubbles et al., *Climate Science Special Report*. The CSSR (p. 127) provided the following summary paragraph regarding scientific confidence in detection of climate change and attribution to human and natural causes:

Detection and attribution studies, climate models, observations, paleoclimate data, and physical understanding lead to *high confidence* (*extremely likely*) that more than half of the observed global mean warming since 1951 was caused by humans, and *high confidence* that internal climate variability played only a minor role (and possibly even a negative contribution) in the observed warming since 1951. The key message and supporting text summarizes extensive evidence documented in the peer-reviewed detection and attribution literature, including in the [Fifth Assessment Report of the Intergovernmental Panel on Climate Change].

confidence has evolved from nearly two centuries of research and assessments.

This chapter describes a chronology (in the Appendix) of 200 years of major scientific statements, selected to represent views at each time, regarding the human and natural contributions to global climate change. The chronology demonstrates how scientific views and confidence in those views evolved over time.

That GHGs, including carbon dioxide (CO_2),[7] water vapor, and other gases, warm the Earth's climate is not a recent concept. The greenhouse effect,[8] as it is sometimes called, was deduced as early as 1827[9] with relatively little dispute since the 19th century among scientists about the role of GHGs: Some level of GHGs in the atmosphere is necessary for maintaining a temperate climate on Earth. Instead, the debate that unfolded involved whether the climate had been warming overall[10] and, if so, to what the changes may be attributable (such as industrial releases of GHGs, volcanoes, solar activity, or other natural variations). (See text box, Human and Natural Influences on Climate.)

---

[7] CO_2 was referred to as carbonic acid in the 1800s, as will appear in some entries in **Table A-1**.

[8] GHGs in the atmosphere allow shortwave solar radiation to pass through to the Earth's surface but block longwave energy (i.e., heat) from re-radiating into space. GHGs are similar to glass that allows sunlight to enter a greenhouse but keeps heat from escaping. An early description is present in T. Sterry Hunt, "On the Earth's Climate in Palæozoic Times." *The London, Edinburgh, and Dublin Philosophical Magazine and Journal of Science*, vol. 26, no. 175 (October 1, 1863):

The late researches of Tyndall on the relation of gases and vapours to radiant heat are important in their bearing upon the temperature of the earth's surface in former geological periods. He has shown that heat, from whatever source, passes through hydrogen, oxygen, and nitrogen gases, or through dry air, with nearly the same facility as through a vacuum.... Gases and some other solid substances, which are readily permeable to light and to solar heat, offer, as is well known, great obstacles to the passage of radiant heat from non-luminous bodies; and Tyndall has recently shown that many colourless vapours and gases have a similar effect, intercepting the heat from such sources, by which they become warmed and in their turn radiate heat.... The diffusion of olefiant gas ... carbonic acid gas ... ozone ... watery vapour present in the air.... *Like a covering of glass, it allows the sun's rays to reach the earth but prevents to a great extent the loss by radiation of the heat thus communicated.* (p. 323, emphasis added)

[9] Jean-Baptiste Joseph Fournier, "Mémoire Sur Les Températures Du Globe Terrestre et Des Espaces Planétaires," *Mémoires de l'Académie Royale Des Sciences de l'Institute de France*, 1827, https://geosci.uchicago.edu/~rtp1/papers/ Fourier1827Trans.pdf.

[10] The focus of this chapter is on attribution of global temperature changes, but much research and many scientific assessments address other metrics of climate, such as precipitation and seasonality.

Chronology of Climate Change Science

As indicated by the information presented in Table A-1, scientists have noted, dating back to early in the 19th century, both human and natural factors potentially influencing climate. As one scholar observed, "by 1900, most of the chief theories of climate change had been proposed, if not yet fully explored."[11] There were a number of contending theories—including changes in solar energy, the Earth's orbital geometry, volcanoes, the geography of continents, and changes in GHGs—in the late 1800s as the quotations in Table A-1 indicate. Well into the 1900s, the state of the science relating CO_2 concentrations in the atmosphere to the Earth's temperature was primarily theoretical inference. Scientists debated whether increases in CO_2 in the atmosphere due to increasing emissions from fossil fuels would lead to further warming.

Since then, a number of factors—including better measurement technologies; development of physics- and empirically based simulation models; more research, review, and revision; and longer series of observations—have improved the foundations of climate science. As a result, scientists have improved quantification of the relationships between observed conditions:

1. Natural and human-related GHG emissions (the latter mostly from fossil-fuel- based energy) to the atmosphere;
2. Increasing GHG concentrations in the atmosphere and changes in other influences on climate (e.g., changes in solar and volcanic activity);
3. Rising global average surface temperature; and
4. Other observed changes in the spatial and temporal patterns of climate.

---

[11] James Rodger Fleming, *Historical Perspectives on Climate Change*, Oxford University Press (2005).

Human and Natural Influences on Climate[12]

There are several ways that climate can be affected by both human and natural causes.

Humans: Climate is affected by changing concentrations in the atmosphere of GHGs such as CO_2, methane, water vapor, chlorofluorocarbons, hydrofluorocarbons, and other gases and aerosols. GHGs trap heat in the atmosphere and warm the atmosphere. Most human-related GHGs are emitted in the production and use of fossil fuels. Though not typically considered GHGs, emissions of polluting sulfur and carbonaceous aerosols also alter the atmosphere's reflection of solar radiation and its absorption of heat. Humans change land cover, affecting both the reflectivity of land and the removal of CO_2 from the atmosphere by vegetation. Higher CO_2 concentrations in the atmosphere fertilize vegetation, increasing removals of CO_2 from the atmosphere when other factors (e.g., nutrient and water availability, temperature, solar insolation, etc.) are not constraining.

Natural: Natural influences on the climate may include changes in solar energy, naturally occurring water vapor and CO_2 in the atmosphere, volcanic aerosol and GHG emissions, and cyclical oscillations in the oceans. These factors impact climate by affecting the amount of solar radiation reaching earth's surface, modulating how much heat is retained within the atmosphere, or changing oceanic and atmospheric temperature circulation. Over scales of centuries to tens of thousands of years, the Earth's rotational wobble and orbit around the sun also affect the pattern of incoming solar radiation and affect climate on geologic time scales (e.g., glacial and inter-glacial periods).

[12] Weubbles et al., *Climate Science Special Report.*

The magnitudes of factors, and therefore their influences on climate, vary over time. With acceleration of human population growth and industrialization since the 19th century, the factors related to human activities have increased relative to those of natural processes. Increased scientific capacity has made climate change increasingly detectable and attributable to the varying influences over the past two centuries.[13]

In the late 1930s, Guy Callendar compiled existing data on atmospheric CO_2 concentrations and regional temperatures.[14] Through imprecise calculations, he showed a correlation between observed increases in both over time. Some scientists considered the correlation merely coincidence. Callendar's calculations provided early quantitative indications of a climate warming as a result of human activity. At the time, however, the relative contribution of human activity compared with natural factors could not be determined.[15] David Keeling later established more consistent and repeatable measurements of atmospheric CO_2 in the 1950s.[16] Keeling's precise measurements provided strong evidence of a connection between increasing human- related CO_2 emissions and the increasing CO_2 concentrations in the atmosphere.

---

[13] Before robust observations of the climate and the factors that influence it became available, scientists theorized how future climate would evolve with, for example, continued increases in fossil fuel emissions, as can be seen in the quotations in **Table A-1**. Since the 1990s, projections of future climate changes rely on formally produced and peer- reviewed scenarios of how different climate forcing factors may evolve, including how economic activities and public policies may alter those factors.

[14] Guy S. Callendar, "The Artificial Production of Carbon Dioxide and Its Influence on Temperature," *Quarterly Journal of the Royal Meteorological Society*, vol. 64, no. 275 (April 1938), pp. 223-40, https://www.eas.ualberta.ca/ jdwilson/EAS372_15/ exams/Callendar_QJRMS1938.pdf.

[15] Substantial scientific research suggests that solar radiation and other natural factors are important in the global temperature increase in the first half of the 20th century. See, among other research, P.A. Stott et al, "Attribution of Twentieth Century Temperature Change to Natural and Anthropogenic Causes," *Climate Dynamics*, vol.17, no. 1 (January 1, 2001), pp. 1-21, https://doi.org/10.1007/PL00007924.

[16] Scripps Institute of Oceanography, "The Early Keeling Curve," http://scrippsco2.ucsd.edu/ history_legacy/early_keeling_curve.

The measurements established a quantitative benchmark for later studies examining the linkage between increasing CO_2 concentrations and rising global temperatures. Keeling's concentration data facilitated additional research on the global carbon cycle, the oceans, and the effects of human activities.

In the middle of the 20th century, scientists (and many in the public) recognized that a general warming of the climate had occurred (Table A-1).[17] This was followed by a 30-year period of relatively flat or decreasing global average temperatures from around 1946 to 1977[18] (Figure 1).

Arguably, the apparent change in trajectory heightened scientific uncertainty about the direction of future climate changes and any human influence on them. It coincided with concerns about "global dimming,"[19] at least in part attributed to sulfur and particulate pollution,[20] which increased rapidly during that period before leveling off around 1980.[21]

Current assessments indicate that the mid-20th century warming hiatus may have been due to a combination of human (GHG, pollution) and natural influences (solar variability, volcanoes).[22]

---

[17] See, for example, Albert Abarbanel and Thorp McClusky, "Is the World Getting Warmer?," *The Saturday Evening Post*, July 1, 1950.

[18] National Academy of Sciences, *Energy and Climate* (Washington, DC: National Academies Press, 1977), https://www.nap.edu/download/12024.

[19] For a relatively recent review of research on the variability of solar radiation reaching the Earth's surface— sometimes called "global dimming" and "brightening" by scientists—and how it may have influenced non-linear changes in 20th century climate, see Martin Wild, "Global Dimming and Brightening: A Review," *Journal of Geophysical Research*, vol. 114, June 27, 2009, https://doi.org/200910.1029/2008JD011470.

[20] According to this theory, global dimming and cooling during this time was probably at least partially influenced by human activities in the form of aerosol pollution, as explained in the article cited above.

[21] See, among others, Martin Wild, "Enlightening Global Dimming and Brightening," *Bulletin of the American Meteorological Society*, vol. 93, no. 1 (July 15, 2011), http://journals.ametsoc.org/doi/pdf/10.1175/BAMS-D-11-00074.1.

[22] N. L. Bindoff et al, "Detection and Attribution of Climate Change: from Global to Regional," chapter 10 in T. F. Stocker et al. (eds.), *Climate Change 2013: The Physical Science Basis: Contribution of Working Group I to the Fifth Assessment Report of the Intergovernmental Panel on Climate Change* (Cambridge: U.K.: Cambridge University Press, 2013), p. 883, http://ipcc.ch/report/ar5/wg1/. This chapter also describes the methods and data for "detection and attribution" of climate change over different time scales.

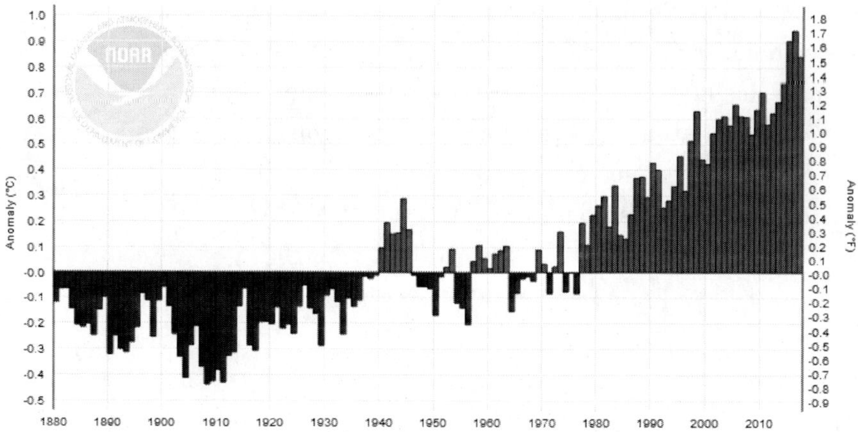

Source: National Oceanic and Atmospheric Administration, National Centers for Environmental Information, "Climate at a Glance: Global Time Series," annual data for 1880-2017, retrieved on January 24, 2018, https://www.ncdc.noaa.gov/cag/time-series/global/globe/land_ocean/ytd/12/1880-2017.

Notes: The temperature "anomaly" is the difference between the measured temperature and the average over the 20th century. Statisticians consider computation of the anomalies to be more reliable than using absolute temperature measurements. For documentation, see https://www.ncdc.noaa.gov/ghcnm/v3.php. Four other research organizations analyze and publish global temperature anomalies. They use different methods and produce similar estimates—particularly regarding trends—with NOAA's typically being slightly below those of the National Aeronautics and Space Administration, for example.

Figure 1. Anomalies of Global Average Surface Temperature, 1880-2017. (in degrees Celsius and Fahrenheit).

As temperatures began to rise again in the late 1970s, authoritative scientific assessments performed by various governmental and nongovernmental institutions, supported by an expanding body of peer-reviewed published research, pointed to an emerging consensus regarding a probable human contribution to climate change, primarily due to increasing GHG emissions.

The relative role of human versus natural influences became more clear in the early 2000s. Longer series of improved observations (e.g., solar radiation, clouds, land cover change), statistical methods, and computational models enabled more robust analyses and comparisons of research methods and results. Major, collaborative, authoritative assessments—U.S. and international—were established to compile, debate, and consider the strengths or weaknesses of scientific analysis regarding climate change in order to inform policymakers.

What Are Scientific Assessments of Climate?

Policymakers, when seeking thoroughly vetted and balanced scientific information, sometimes commission scientific assessments. On climate change, periodic assessments have been mandated under domestic law (e.g., the Global Change Research Act of 1990, P.L. 101-606) and by international agreements (United Nations General Assembly Resolution 43/53 of 6 December 1988). According to the Office of Management and Budget, a scientific assessment is

> an evaluation of a body of scientific or technical knowledge, which typically synthesizes multiple factual inputs, data, models, assumptions, and/or applies best professional judgment to bridge uncertainties in the available information. These assessments include, but are not limited to, state-of-science reports; technology assessments; weight-of-evidence analyses; meta- analyses....[23]

The National Academy of Sciences (NAS), the Intergovernmental Panel on Climate Change (IPCC), and the United States Global Change Research Program (USGCRP) have organized scientific climate assessments involving hundreds to thousands of individual scientists and subject to expert peer reviews and, for the IPCC and USGCRP, public peer reviews.

[23] U.S. Office of Management and Budget, "Final Information Quality Bulletin for Peer Review," 70 *Federal Register* 2667, January 14, 2005, https://www.gpo.gov/fdsys/pkg/FR-2005-01-14/pdf/05-769.pdf.

Table A-2 contains the relevant conclusions regarding human and natural influences on climate change from the major assessments conducted, with those of

- NAS beginning in 1977,[24]
- IPCC beginning in 1990, and
- the USGCRP beginning in 2001.

These inclusive assessments underpinned growing scientific confidence that human activities were likely the major cause of the observed global warming since the mid-20th century.

CONCLUSION

Many factors have contributed to increased scientific confidence in quantifying the human and natural contributions to climate change. Longer records of observational data have provided more evidence of the concordance between higher GHG levels and temperature increases. Satellites have provided important observations of temperatures; atmospheric pollution; and land, snow, and ice cover beginning in the late 1970s. Additionally, improved scientific understanding of atmospheric physics, together with vastly more powerful computers, has led to climate models that better simulate atmospheric and oceanic conditions. Uncertainties in the models remain on how they simulate the effects of clouds, for example, and the model simulations are at smaller scales of space and time. Despite these uncertainties, current climate scientific assessment states high confidence (extremely likely) that human influence is the dominant cause of the observed warming over the past half-century. While the near consensus has developed relatively recently, it has evolved based on increasing confidence through research on scientific concepts

---

[24] NAS is a private, nonprofit society of distinguished scholars. It was established by Congress in statute in 1863 and is charged with providing independent, objective advice to the nation on matters related to science and technology.

established as early as 200 years ago. Future climate outcomes depend on many additional factors, such as the future rates and character of socio-economic development and efforts to curtail the growth of GHG emissions.

APPENDIX. CHRONOLOGY OF SCIENTIFIC STATEMENTS ON ATTRIBUTION OF GLOBAL CLIMATE CHANGE

This appendix contains bibliographic references and quotations regarding scientific understanding of global climate change and the influence of CO_2, other GHGs, and natural factors on observed and prospective global climate. Because the capacities and methods of science have changed markedly over the past 200 years, the references appear in two tables representing selected scientific literature and national or international scientific assessments.

Table A-1 presents representative statements excerpted from key scientific literature from 1827 to 1987 regarding human-related and other contributions to climate change. Sources include selected, widely cited academic papers, government reports, and NAS reports. The table's selections largely precede the establishment of broadly inclusive scientific assessments to compile and assess the weight of scientific evidence. For the period up to 1987, CRS selected key academic scientific papers and reports that were influential to scientific contemporaries during and after their respective times.

Scientific assessments began in the mid-20th century to more systematically and inclusively evaluate the full body of scientific literature on specific topics.

Table A-2 compiles the conclusions pertinent to this chapter from major U.S. and international scientific assessments, beginning in 1977, that address the human contribution to global climate change. The assessments have been produced by the USGCRP, NAS, and IPCC.[25]

---

[25] Individual scientists, and some organizations, may not agree with the conclusions of major scientific assessments. CRS is not aware of any national or international climate-related scientific body that dissented from the recent assessments quoted in this chapter.

Table A-1. Select Scientific Statements Regarding the Influences on Global Climate of Human Activities and Natural Factors, 1827-1987

Year	Title	Key Quotation(s)	Reference
1827	Mémoire sur les Températures du Globe Terrestre et des Espaces Planétaires [On the Temperatures of the Terrestrial Sphere and Interplanetary Space]	"The establishment and progress of human societies, and the action of natural forces, can notably change the state of the ground surface over vast regions, as well as the distribution of waters and the great movements of the air. Such effects have the ability to make the mean degree of heat vary over the course of several centuries."	Jean-Baptiste Joseph Fournier, "Mémoire Sur Les Températures Du Globe Terrestre et Des Espaces Planétaires," *Mémoires de l'Académie Royale Des Sciences de l'Institute de France*, 1827, https://geosci.uchicago.edu/~rtp1/ papers/Fourier1827Trans.pdf.
1856	Circumstances Affecting the Heat of the Sun's Rays	"An atmosphere of [carbonic acid] would give to our earth a high temperature; and if as some suppose, at one period of its history the air had mixed with it a larger proportion than at present, an increased temperature from its own action as well as from increased weight must have necessarily resulted."	Eunice Foote, "Circumstances Affecting the Heat of the Sun's Rays," *American Journal of Science and Arts*, vol. 22 (1856), pp. 382-83, https://books.google.com/ books?id=6xhFAQAAMAAJ&pg= PA382#v=onepage&q&f=false.
1896	On the Influence of Carbonic Acid in the Air upon the Temperature of the Ground	"If the quantity of carbonic acid increases in geometric progression, the augmentation of the temperature will increase nearly in arithmetic progression." "This quantity of carbonic acid, which is supplied to the atmosphere chiefly by modern industry, may be regarded as completely compensating the quantity of carbonic acid that is consumed in the formation of limestone (or other mineral carbonates) by the weathering or decomposition of silicates." "We must regard volcanic exhalations as the chief source of carbonic acid for the atmosphere."	Svante Arrhenius, "On the Influence of Carbonic Acid in the Air upon the Temperature of the Ground," *Philosophical Magazine and Journal of Science*, vol. 41, series 5 (April 1896), pp. 237-76, http://www.trunity.net/files/108501_108600/108531/ arrhenius1896_greenhouse-effect.pdf.
1897	A Group of Hypotheses Bearing on Climatic Changes	"A comparison of early with later life, stripped of theoretical presumptions, does not seem to me clearly to imply any great difference in the content of carbon dioxide." "For it is almost axiomatic to say that climatic changes would attend changes in the constitution of the atmosphere. I assume that atmospheric poverty, especially in the critical item of carbon dioxide, is correlated with low temperature."	Thomas C. Chamberlin, "A Group of Hypotheses Bearing on Climatic Changes," *Journal of Geology*, vol. 5, no. 7 (1897), pp. 653-683, https://www.jstor.org/stable/pdf/ 30054630.pdf?refreqid= excelsior%3Aa77b10a39bbbe564c432a9 204d719cec.
1898	Monthly Weather Review and Annual Summary	"We have therefore an alternation of cold and warm periods due to the interaction of elevation [of the Earth's surface] and carbon dioxide [in the atmosphere].... [W]e hope eventually to be able to deduce the resulting climatic peculiarities, and to show that very slight changes in	U.S. Department of Agriculture, Weather Bureau, *Monthly Weather Review and Annual Summary*, July 1898, http://congressional.proquest.com/ congressional/result/pqpresultpage.gispdfhitspanel.pdflink/

Table A-1. (Continued)

Year	Title	Key Quotation(s)	Reference
		oceans and continents have produced all the variations of geological climate, and that little or nothing need be hypothecated as to the variations of solar heat, of atmospheric gases, of terrestrial latitudes, or the many other climatological elements."	$2fapp-s3$2fjlh-content$2fPT4-A2906-8$2fPT4-A2906-8_From_1_to_752.pdf/ entitlementkeys=1234%7Capp- gis%7Cexecutive-branch%7Ca2906-8.
1911	USGS: The Data of Geochemistry	"In a thousand years, then, if the rate were constant and no disturbing factors interfered, the amount of CO_2 in the atmosphere would be doubled. If we take into account the combustion of fuels other than coal and the large additions to the atmosphere from the sources previously mentioned the result becomes still more startling. Were there no counterbalancing of this increase in atmospheric carbon, animal life would soon become impossible upon our planet." "Whether the theory of Arrhenius is in harmony with the facts of historical geology—that is, whether periods of volcanic activity [releasing carbon dioxide] have coincided with warmer climates, and a slackening of activity with lowering of temperatures is also in dispute. Apparently, the controversy is not yet ended." "Water vapor, then, is the chief agent in the atmospheric regulation of climate and to this conclusion; The climatic conditions may vary as Chamberlin claims, but the relative dryness or wetness of the atmosphere may be the true cause of fluctuating temperatures, rather than the carbon dioxide."	Frank Wigglesworth Clark, U.S. Geological Survey, "The Data of Geochemistry", *Bulletin 491, 2nd ed., U.S. House of Representatives*, 1911, pp. 47, 135, https://congressional.proquest.com/ congressional/result/ pqpresultpage.gispdfhitspanel.pdflink/$2fapp-bin$2fgis-serialset$2f0$2fd$2f4$2fa$2f5997_hdoc 1499_from_1_to_779.pdf/entitlementkeys=1234%7Capp-gis%7Cserialset%7C5997_h.doc.1499.
1939	The Composition of the Atmosphere Through the Ages	"It is a commonplace that man is able to speed up the processes of Nature, and he has now plunged heavily into her slow-moving carbon cycle by throwing some 9,000 tons of carbon dioxide into the air each minute. This great stream of gas results from the combustion of fossil carbon (coal, oil, peat, etc.), and it appears to be much greater than the natural rate of fixation." "From the best laboratory observations it appears that the principal result of increasing atmospheric carbon dioxide, apart from a slight speeding up of rock weathering and plant growth, would be a gradual increase in the mean temperature of the colder regions of the earth."	Guy S. Callendar, "The Composition of the Atmosphere Through the Ages," *Meterological Magazine*, vol. 74, no. 878 (March 1939), pp. 33-39, https:// www.google.com/url?sa=t&rct= j&q=&esrc=s&source=web&cd=1&ved= 0ahUKEwjBjtHfoP3WAhXB4SYKHYGF A9UQFggmMAA&url= https%3A%2F%2Fdigital.nmla.metoffice.g ov.uk%2Fdownload%2Ffile%2Fsdb%253 AdigitalFile%257Cedc8aafc-a27a-4afe- bd29-619b052bd199%2F&usg= AOvVaw0R8QW-XaTql-iGP-cJjF11https://www.google.com/url?sa=t& rct=j&q=&esrc=s&source=web&cd=1& ved=

Year	Title	Key Quotation(s)	Reference
			0ahUKEwjBjtHfoP3WAhXB4SYKHYGF A9UQFggmMAA&url= https%3A%2F%2Fdigital.nmla.metoffice.gov.uk%2Fdownload%2Ffile%2Fsdb%253 AdigitalFile%257Cedc8aafc-a27a-4afe- bd29-619b052bd199%2F&usg= AOvVaw0R3QW-XaTql-iGP-cJjF11
1948	Circular Causal Systems in Ecology	"Callendar (1940), considering the best data available since 1866, concludes that during the present century there has been an increase of the order of 10 per cent in the CO_2 content of the atmosphere. This he attributes to the modern industrial combustion of fuel.... It seems far more likely that the observed increment in the carbon dioxide of air at low levels in both Europe and eastern North America is due to changes in the biological mechanisms of the cycle rather than to an increase in industrial output.... The self-regulating mechanisms of the carbon cycle can cope with the present influx of carbon of fossil origin."	George Evelyn Hutchinson, "Circular Causal Systems in Ecology," *Annals of the New York Academy of Sciences*, vol. 50, (1948), pp. 221-246, http:// people.wku.edu/charles.smith/ biogeog/HUTC1948.htm.
1966	U.S. House of Representatives: Environmental Pollution: A Challenge to Science and Technology	"Carbon dioxide accumulations from the burning of gas, petroleum, and coal change the nature of the atmosphere. Weather patterns can be altered purposefully or accidentally by human activity. These powerful forces have only come about recently and are not well understood. As a consequence, in many risk-benefit questions, the magnitude of the risk is relatively unknown." "By the year 2000 at present and projected rate of fossil fuel consumption, the carbon dioxide level in the air will cause a significant rise in the air temperature."	Rep. George P. Miller, "Environmental Pollution: A Challenge to Science and Technology," *Report of the Subcommittee on Science, Research, and Development. Committee on Science and Astronautics, U.S. House of Representatives*, 1966, pp. 3, 43, 263, https://congressional.proquest.com/ congressional/result/ pqpresultpage.gispdfhitspanel.pdflink/$2fapp-bin$2fgis-congresearch$2f6$2f7$2f2$2f9$2fcmp-1966-sah-0009_from_1_to_67.pdf/entitlementkeys=1234%7Capp-gis%7Ccongresearch%7Ccmp-1966-sah-0009.
1970	Man-Made Climatic Changes	"It appears that on the local scale man- made influences on climate are substantial but that on the global scale natural forces still prevail.... The potential for anthropogenic changes of climate on a larger and even a global scale is real.... In my opinion, man-made aerosols, because of their optical properties and possible influences on cloud and precipitation processes constitute a more acute problem than CO_2.... Over longer intervals, energy added to the atmosphere by heat rejection and CO_2 absorption remain matters of concern."	Helmut E. Landsberg, "Man-Made Climatic Changes," *Science*, vol. 170, no. 3964 (December 18, 1970), http://www.jstor.org/stable/pdf/1730486.pdf?refreqid= excelsior%3Ae79d4c4c72387997406c413dafdebbaf.
1972	NSF: Patterns and Perspectives in Environmental Science	"Judging from the record of the past interglacial ages, the present time of high temperatures should be drawing to an end, to be followed by a long period of considerably colder temperatures leading into the next	National Science Board, National Science Foundation, "Patterns and Perspectives in Environmental Science," 1972, https://archive.org/details/patternsperspect00nati.

Table A-1. (Continued)

Year	Title	Key Quotation(s)	Reference
		glacial age some 20,000 years from now. However, it is possible, or even likely, that human interference has already altered the environment so much that the climatic pattern of the near future will follow a different path. For instance, widespread deforestation in recent centuries, especially in Europe and North America, together with increased atmospheric opacity due to man-made dust storms and industrial wastes, should have increased the earth's reflectivity. At the same time, increasing concentration of industrial carbon dioxide in the atmosphere should lead to a temperature increase by absorption of infrared radiation from the earth's surface. When these human factors are added to such other natural factors as volcanic eruptions, changes in solar activity, and resonances within the hydro-atmosphere, their effect can only be estimated in terms of direction, not of amount."	
1974	NSF: Science and the Challenges Ahead	"Human activity may be involved on an even broader scale in changing the global climate.... The cause of the cooling trend is not known with certainty. But there is increasing concern that man himself may be implicated, not only in the recent cooling trend but also in the warming temperatures over the last century.... By the middle of this century, the cooling effect of the dust particles more than compensated for the warming effect of the carbon dioxide, and world temperature began to fall."	National Science Board, National Science Foundation, "Science and the Challenges Ahead," 1974, https://archive.org/details/ sciencechallenge00nati.
1975	NAS: Understanding Climate Change: A Program for Action	"While the natural variations of climate have been larger than those that may have been induced by human activities during the past century, the rapidity with which human impacts threaten to grow in the future, and increasingly to disturb the natural course of events, is a matter of concern. These impacts include man's changes of the atmospheric composition and his direct interference with factors controlling the all important heat balance.... Of the two forms of pollution [CO_2 and aerosols], the carbon dioxide increase is probably the more influential at the present time in changing temperatures near the earth's surface. If both the CO_2 and particulate inputs to the atmosphere grow at equal rates in the future, the widely differing atmospheric residence times of the two pollutants means that the particulate effect will grow in importance relative to that of CO_2. There are other possible impacts of human activities that should be considered in projecting future climates."	National Academy of Sciences. *Understanding Climate Change: A Program for Action* (Washington, DC: National Academies Press, 1975), https://ia801806.us.archive.org/7/items/ understandingcli00unit/ understandingcli00unit.pdf.

Year	Title	Key Quotation(s)	Reference
1983	World Climate Research Programme: Report of the Fifth Session of the Joint Scientific Committee	"The existing observational evidence does not allow concluding with confidence that the observed variations are caused specifically by the increasing concentration of atmospheric carbon dioxide. On the other hand, the surface warming indicated by the record of the global mean temperature is not inconsistent with the computed effect of the observed increase of atmospheric CO_2 since the beginning of the industrial revolution."	International Council of Scientific Unions, World Meteorological Organization, *Report of the Fifth Session of the Joint Scientific Committee (Hangzhou, China, 12-17 March 1984)*, September 1984, https://library.wmo.int/pmb_ged/wmo-td_1.pdf.
1985	Villach Conference Statement	"Based on analyses of observational data, the estimated increase in global mean temperature during the last one hundred years of between 0.3 and 0.7°C is consistent with the projected temperature increase attributable to the observed increase in CO2 and other greenhouse gases, although it cannot be ascribed in a scientifically rigorous manner to these factors alone."	World Meteorological Organisation, *Report of the International Conference on the assessment of the role of carbon dioxide and of other greenhouse gases in climate variations and associated impacts,* Villach, Austria, October 9-15, 1985, WMO No. 661, 1986, https://web.archive.org/web/20131121040937/http:// www.scopenvironment.org/downloadpubs/scope29/statement.html.
1987	Mankind's Impact on Climate: The Evolution of an Awareness	"There is now a strong consensus that the observed increase in the atmospheric concentrations of carbon dioxide and other infrared-absorbing trace gases is indeed warming the earth, and that this change is caused by mankind."	William W. Kellogg, "Mankind's Impact on Climate: The Evolution of an Awareness," *Climatic Change*, vol. 10 (1987), pp. 113-36, https://link.springer.com/content/pdf/10.1007%2FBF00140251.pdf.

Source: Compiled by CRS.

Table A-2. Excerpted Statements on the Human Contribution to Global Climate in Scientific Assessments of the NAS, the IPCC, and the USGCRP, 1977-2017

Year	Title	Key Quotation(s)	Reference
1977	NAS: Energy and Climate: Studies in Geophysics	"In the first half of the twentieth century ... a general warming of the earth occurred.... There is considerable evidence that, between the 1940s and about 1970, the climatic changes of the earlier part of this century had tended to undergo a reversal. Temperatures had mostly fallen." "Today, we find ourselves uneasy because through our use of energy we may be significantly disturbing the natural climate system. This uneasiness is justified."	National Academy of Sciences, *Energy and Climate* (Washington, DC: National Academies Press, 1977), https://www.nap.edu/download/12024.
1979	NAS: Carbon Dioxide and Climate: A Scientific Assessment	"None of the model calculations predicts negligible warming. The primary effect of an increase of CO_2 is to cause more absorption of thermal radiation from the earth's surface and thus to increase the air temperature in the troposphere." "We conclude that the predictions of CO_2-induced climate changes made with the various models examined are basically consistent and mutually supporting."	National Academy of Sciences, "Carbon Dioxide and Climate: A Scientific Assessment," Report of an Ad Hoc Study Group on Carbon Dioxide and Climate, Woods Hole, Massachusetts, July 23-27, 1979, to the Climate Research Board, Assembly of Mathematical and Physical Sciences, National Research Council, https://www.nap.edu/download/19856.
1990	IPCC: Climate Change: The IPCC Scientific Assessment - 1st Report	"The size of the warming over the last century is broadly consistent with the predictions of climate models but is also of the same magnitude as natural climate variability. If the sole cause of the observed warming were the human made greenhouse effect, then the implied climate sensitivity would be near the lower end of the range inferred from the models. The observed increase could be largely due to natural variability, alternatively this variability and other man-made factors could have offset a still larger man-made greenhouse warming. The unequivocal detection of the enhanced greenhouse effect from observations is not likely for a decade or more, when the commitment to future climate change will then be considerably larger than it is today."	J. T. Houghton et al., *Climate Change: The IPCC Scientific Assessment.* Report prepared for Intergovernmental Panel on Climate Change by Working Group I (Cambridge, U.K.: Cambridge University Press, 1990), https://www.ipcc.ch/ipccreports/far/wg_I/ipcc_far_wg_I_full_report.pdf.
1994	NAS: Solar Influences on Global Change	"Taken collectively, the above evidence, although circumstantial, does suggest that solar variability could influence future global change.... Lack of knowledge of solar influences will limit the certainty with which anthropogenic climate change can be detected. But it is unlikely that solar influences on global change will be comparable to the expected anthropogenic influences."	National Academy of Sciences, *Solar Influences on Global Change* (Washington, DC: National Academies Press, 1994), https://www.nap.edu/download/4778.
1995	IPCC Second Assessment: Climate	"Carbon dioxide remains the most important contributor to anthropogenic forcing of climate change; projections of future global mean temperature	B. Bolin et al., *Climate Change 1995: A report of the Intergovernmental Panel on Climate Change* (Cambridge, U.K.:

Year	Title	Key Quotation(s)	Reference
	Change 1995	change and sea level rise confirm the potential for human activities to alter the Earth's climate to an extent unprecedented in human history.... [O]bservations suggest a 'discernible human influence on global climate.'" "Most of these studies have detected a significant change and show that the observed warming trend is unlikely to be entirely natural in origin."	Cambridge University Press, 1995), http://www.ipcc.ch/ipccreports/sar/ wg_I/ipcc_sar_wg_I_full_report.pdf.
2001	USGCRP: 1st National Climate Assessment	"The observed magnitude, pattern, and timing of the global warming indicate that the rising concentrations of CO_2 and other greenhouse gases caused by human activities are contributing significantly to the recent warming."	National Assessment Synthesis Team, *Climate Change Impacts on the United States: The Potential Consequences of Climate Variability and Change*, Report for the USGCRP (Cambridge, U.K.: Cambridge University Press, 2001), https://data.globalchange.gov/assets/e9/97/436129058f2107f4925aeec13ed8/nca-2000-foundation-report.pdf.
2001	NAS: Climate Change Science: An Analysis of Some Key Questions	"The changes observed over the last several decades are likely mostly due to human activities, but we cannot rule out that some significant part of these changes is also a reflection of natural variability."	National Academy of Sciences, *Climate Change Science: An Analysis of Some Key Questions* (Washington, DC: National Academy Press, 2001), https://www.nap.edu/download/10139.
2001	IPCC: Climate Change 2001: The IPCC Third Assessment Report	"There is new and stronger evidence that most of the warming observed over the last 50 years is attributable to human activities."	R. T. Watson et al. (eds.), *Climate Change 2001: Synthesis Report*, Contribution of Working Groups I, II, and III to the Third Assessment Report of the Intergovernmental Panel on Climate Change (Cambridge, U.K.: Cambridge University Press, 2001), https://www.ess.uci.edu/researchgrp/prather/files/2001IPCC_SyR- Watson.pdf.
2007	IPCC: The Fourth Assessment Report: Climate Change	"Most of the observed increase in global average temperatures since the mid-20th century is very likely [> 90%] due to the observed increase in anthropogenic GHG concentrations."	R. K. Pachuri et al. (eds.), *Climate Change 2007: Synthesis Report*, Contribution of Working Groups I, II, and III to the Fourth Assessment Report of the Intergovernmental Panel on Climate Change (Geneva, Switzerland: IPCC Secretariat, 2007), http://www.slvwd.com/agendas/Full/2007/06-07-07/Item%2010b.pdf.
2009	USGCRP: 2nd National Climate Assessment	"The global warming of the past 50 years is due primarily to human-induced increases in heat trapping gasses.... Recent scientific assessments find that most of the warming of the Earth's surface over the past 50 years has been caused by human activities."	Thomas R. Karl et al. (eds.), *Global Climate Change Impacts in the United States* (Cambridge, UK: Cambridge University Press, 2009), https://downloads.globalchange.gov/usimpacts/pdfs/climate-impacts- report.pdf.
2010	NAS: Advancing the Science of Climate Change	"Climate change is occurring, is caused largely by human activities, and poses significant risks for—and in many cases is already affecting—a broad range of human and natural systems."	National Academy of Sciences, *Advancing the Science of Climate Change*, America's Climate Choices: Panel on Advancing the Science of Climate Change (Washington, DC: National Academy Press, 2010), https://www.nap.edu/download/12782.

Table A-2. (Continued)

Year	Title	Key Quotation(s)	Reference
2012	NAS: Climate Change: Evidence, Impacts, and Choices	"The overwhelming majority of climate scientists agree that human activities, especially the burning of fossil fuels (coal, oil, and gas), are responsible for most of the climate change currently being observed."	National Academy of Sciences, *Climate Change: Evidence, Impacts, and Choices*, 2012, http://nas-sites.org/americasclimatechoices/files/2012/06/19014_cvtx_R1.pdf
2014	NAS: Climate Change: Evidence and Causes	"Rigorous analysis of all data and lines of evidence shows that most of the observed global warming over the past 50 years or so cannot be explained by natural causes and instead requires a significant role for the influence of human activities."	National Academy of Sciences, *Climate Change: Evidence and Causes*, 2014, http://dels.nas.edu/resources/static- assets/exec-office-other/climate-change- full.pdf
2014	IPCC: The Fifth Assessment Report: Climate Change 2014	"[Anthropogenic GHG emissions] effects, together with those of other anthropogenic drivers, have been detected throughout the climate system and are extremely likely [>95%] to have been the dominant cause of the observed warming since the mid-20th century."	R. K. Pachuri et al. (eds.), *Climate Change 2014: Synthesis Report*, Contribution of Working Groups I, II, and III to the Fifth Assessment Report of the Intergovernmental Panel on Climate Change (Geneva, Switzerland: IPCC Secretariat, 2014), http://www.ipcc.ch/pdf/assessment- report/ar5/syr/SYR_AR5_FINAL_full_wcover.pdf.
2014	USGCRP: 3rd National Climate Assessment	"The majority of the warming at the global scale over the past 50 years can only be explained by the effects of human influences, especially the emissions from burning fossil fuels (coal, oil, and natural gas) and from deforestation."	Jerry M. Melillo et al., (eds.), *Highlights of Climate Change Impacts in the United States: The Third National Climate Assessment*, USGCRP (Washington, DC: U.S. Government Printing Office, 2014), http://www.globalchange.gov/sites/globalchange/files/ NCA3_Highlights_LowRes-small-FINAL_posting.pdf.
2017	USGCRP: Climate Science Special Report: Fourth National Climate Assessment	"It is extremely likely [> 95%] that human influence has been the dominant cause of the observed warming since the mid-20th century. For the warming over the last century, there is no convincing alternative explanation supported by the extent of the observational evidence."	Donald J. Weubbles et al., *Climate Science Special Report: Fourth National Climate Assessment, Volume I*, USGCRP, 2017, https://science2017.globalchange.gov/downloads/CSSR2017_FullReport.pdf.

Source: Compiled by CRS.

Notes: The documents cited reflect the evolution of select aspects of climate change science. This list is not comprehensive of, and is not intended to highlight, all aspects of the scientific efforts in climate research.

Table A-3. Select Scientific Statements Regarding the Influences on Global Climate of Human Activities and Natural Factors, 1827-1987

Year	Title	Key Quotation(s)	Reference
1827	Mémoire sur les Températures du Globe Terrestre et des Espaces Planétaires [On the Temperatures of the Terrestrial Sphere and Interplanetary Space]	"The establishment and progress of human societies, and the action of natural forces, can notably change the state of the ground surface over vast regions, as well as the distribution of waters and the great movements of the air. Such effects have the ability to make the mean degree of heat vary over the course of several centuries."	Jean-Baptiste Joseph Fournier, "Mémoire Sur Les Températures Du Globe Terrestre et Des Espaces Planétaires," *Mémoires de l'Académie Royale Des Sciences de l'Institute de France*, 1827, https://geosci.uchicago.edu/ ~rtp1/papers/Fourier1827Trans.pdf.
1856	Circumstances Affecting the Heat of the Sun's Rays	"An atmosphere of [carbonic acid] would give to our earth a high temperature; and if as some suppose, at one period of its history the air had mixed with it a larger proportion than at present, an increased temperature from its own action as well as from increased weight must have necessarily resulted."	Eunice Foote, "Circumstances Affecting the Heat of the Sun's Rays," *American Journal of Science and Arts*, vol. 22 (1856), pp. 382-83, https://books.google.com/books?id=6xhFAQAAMAAJ& pg=PA382#v=onepage&q&f=false.
1896	On the Influence of Carbonic Acid in the Air upon the Temperature of the Ground	"If the quantity of carbonic acid increases in geometric progression, the augmentation of the temperature will increase nearly in arithmetic progression." "This quantity of carbonic acid, which is supplied to the atmosphere chiefly by modern industry, may be regarded as completely compensating the quantity of carbonic acid that is consumed in the formation of limestone (or other mineral carbonates) by the weathering or decomposition of silicates." "We must regard volcanic exhalations as the chief source of carbonic acid for the atmosphere."	Svante Arrhenius, "On the Influence of Carbonic Acid in the Air upon the Temperature of the Ground," *Philosophical Magazine and Journal of Science*, vol. 41, series 5 (April 1896), pp. 237-76, http://www.trunity.net/files/108501_108600/ 108531/arrhenius1896_greenhouse- effect.pdf.
1897	A Group of Hypotheses Bearing on Climatic Changes	"A comparison of early with later life, stripped of theoretical presumptions, does not seem to me clearly to imply any great difference in the content of carbon dioxide." "For it is almost axiomatic to say that climatic changes would attend changes in the constitution of the atmosphere. I assume that atmospheric poverty, especially in the critical item of carbon dioxide, is correlated with low temperature."	Thomas C. Chamberlin, "A Group of Hypotheses Bearing on Climatic Changes," *Journal of Geology*, vol. 5, no. 7 (1897), pp. 653-683, https://www.jstor.org/stable/pdf/ 30054630.pdf?refreqid=excelsior%3Aa77b10a 39bbbe564c432a9204d719cec.
1898	Monthly Weather Review and Annual Summary	"We have therefore an alternation of cold and warm periods due to the interaction of elevation [of the Earth's surface] and carbon dioxide [in the atmosphere].... [W]e hope eventually to be able to deduce the resulting climatic peculiarities, and to show that very slight changes in oceans and continents have produced all the variations of geological climate, and that little or nothing need	U.S. Department of Agriculture, Weather Bureau, *Monthly Weather Review and Annual Summary*, July 1898, http://congressional.proquest.com/ congressional/result/pqpresultpage. gispdfhits panel.pdflink/$2fapp-s3$2fjlh-content$2fPT4- A2906-

Table A-3. (Continued)

Year	Title	Key Quotation(s)	Reference
		be hypothecated as to the variations of solar heat, of atmospheric gases, of terrestrial latitudes, or the many other climatological elements."	8$2fPT4-A2906-8_From_1_to_752.pdf/ entitlementkeys=1234%7Capp- gis%7Cexecutive-branch%7Ca2906-8.
1911	USGS: The Data of Geochemistry	"In a thousand years, then, if the rate were constant and no disturbing factors interfered, the amount of CO_2 in the atmosphere would be doubled. If we take into account the combustion of fuels other than coal and the large additions to the atmosphere from the sources previously mentioned the result becomes still more startling. Were there no counterbalancing of this increase in atmospheric carbon, animal life would soon become impossible upon our planet." "Whether the theory of Arrhenius is in harmony with the facts of historical geology—that is, whether periods of volcanic activity [releasing carbon dioxide] have coincided with warmer climates, and a slackening of activity with lowering of temperatures is also in dispute. Apparently, the controversy is not yet ended." "Water vapor, then, is the chief agent in the atmospheric regulation of climate and to this conclusion; The climatic conditions may vary as Chamberlin claims, but the relative dryness or wetness of the atmosphere may be the true cause of fluctuating temperatures, rather than the carbon dioxide."	Frank Wigglesworth Clark, U.S. Geological Survey, "The Data of Geochemistry", *Bulletin 491, 2nd ed., U.S. House of Representatives*, 1911, pp. 47, 135, https:// congressional. proquest.com/ congressional/result/pqpresult page.gispdfhitspanel.pdflink/$2fapp-bin$2fgis-serialset$2f0$2fd$2f4$2fa$2f5997_hdoc 1499_from_1_to_779.pdf/entitlementkeys= 1234%7Capp-gis%7Cserialset%7 C5997_h.doc.1499.
1939	The Composition of the Atmosphere Through the Ages	"It is a commonplace that man is able to speed up the processes of Nature, and he has now plunged heavily into her slow-moving carbon cycle by throwing some 9,000 tons of carbon dioxide into the air each minute. This great stream of gas results from the combustion of fossil carbon (coal, oil, peat, etc.), and it appears to be much greater than the natural rate of fixation." "From the best laboratory observations it appears that the principal result of increasing atmospheric carbon dioxide, apart from a slight speeding up of rock weathering and plant growth, would be a gradual increase in the mean temperature of the colder regions of the earth."	Guy S. Callendar, "The Composition of the Atmosphere Through the Ages," *Meterological Magazine*, vol. 74, no. 878 (March 1939), pp. 33-39, https://www.google.com/url?sa=t&rct=j&q=& esrc=s&source=web&cd=1&ved= 0ahUKEwjBjtHfoP3WAhXB4SYK HYGFA9UQFggmMAA&url=https%3A %2F%2Fdigital.nmla. metoffice.gov.uk %2Fdownload%2Ffile%2Fsdb%253AdigitalFile %257Cedc8aafc-a27a-4afe-bd29-619b052 bd199%2F&usg=AOvVaw0R8QW-XaTql- iGP-cJjF11https://www.google.com/url?sa=t& rct=j&q=&esrc=s&source=web&cd=1&ved= 0ahUKEwjBjtHfoP3WAhXB4SYKHYGFA 9UQFggmMAA&url=https%3A%2F%2 Fdigital.nmla. metoffice.gov.uk%2Fdownload %2Ffile%2Fsdb%253AdigitalFile%257Cedc

Year	Title	Key Quotation(s)	Reference
			8aafc-a27a-4afe-bd29-619b052bd199%2F& usg=AOvVaw0R8QW-XaTql-iGP-cJjF11
1948	Circular Causal Systems in Ecology	"Callendar (1940), considering the best data available since 1866, concludes that during the present century there has been an increase of the order of 10 per cent in the CO_2 content of the atmosphere. This he attributes to the modern industrial combustion of fuel.... It seems far more likely that the observed increment in the carbon dioxide of air at low levels in both Europe and eastern North America is due to changes in the biological mechanisms of the cycle rather than to an increase in industrial output.... The self-regulating mechanisms of the carbon cycle can cope with the present influx of carbon of fossil origin."	George Evelyn Hutchinson, "Circular Causal Systems in Ecology," *Annals of the New York Academy of Sciences*, vol. 50, (1948), pp. 221-246, http://people. wku.edu/charles.smith/ biogeog/HUTC1948.htm.
1966	U.S. House of Representatives: Environmental Pollution: A Challenge to Science and Technology	"Carbon dioxide accumulations from the burning of gas, petroleum, and coal change the nature of the atmosphere. Weather patterns can be altered purposefully or accidentally by human activity. These powerful forces have only come about recently and are not well understood. As a consequence, in many risk-benefit questions, the magnitude of the risk is relatively unknown." "By the year 2000 at present and projected rate of fossil fuel consumption, the carbon dioxide level in the air will cause a significant rise in the air temperature."	Rep. George P. Miller, "Environmental Pollution: A Challenge to Science and Technology," *Report of the Subcommittee on Science, Research, and Development. Committee on Science and Astronautics, U.S. House of Representatives*, 1966, pp. 3, 43, 263, https://congressional.proquest.com/ congressional/result/pqpresultpage. gispdfhitspanel. pdflink/$2fapp-bin$2fgis-congresearch$2f6$2f7$2f2$2f9$2fcmp-1966- sah-0009_from_1_to_67.pdf/ entitlementkeys=1234%7Capp-gis%7 Ccongresearch%7Ccmp-1966-sah-0009.
1970	Man-Made Climatic Changes	"It appears that on the local scale man-made influences on climate are substantial but that on the global scale natural forces still prevail.... The potential for anthropogenic changes of climate on a larger and even a global scale is real.... In my opinion, man- made aerosols, because of their optical properties and possible influences on cloud and precipitation processes constitute a more acute problem than CO_2.... Over longer intervals, energy added to the atmosphere by heat rejection and CO_2 absorption remain matters of concern."	Helmut E. Landsberg, "Man-Made Climatic Changes," *Science*, vol. 170, no. 3964 (December 18, 1970), http://www.jstor.org/ stable/pdf/1730486.pdf?refreqid=excelsior %3Ae79d4c4c72387997406c413dafdebbaf.
1972	NSF: Patterns and Perspectives in Environmental Science	"Judging from the record of the past interglacial ages, the present time of high temperatures should be drawing to an end, to be followed by a long period of considerably colder temperatures leading into the next glacial age some 20,000 years from now. However, it is possible, or even likely, that human interference has already altered the environment so much that the climatic pattern of the near future will follow a different path. For instance, widespread deforestation in recent centuries, especially in Europe and North America, together with increased atmospheric opacity due to man- made dust storms and industrial wastes, should have increased the earth's reflectivity. At the same time,	National Science Board, National Science Foundation, "Patterns and Perspectives in Environmental Science," 1972, https://archive.org/details/patternsperspect 00nati.

Table A-3. (Continued)

Year	Title	Key Quotation(s)	Reference
		increasing concentration of industrial carbon dioxide in the atmosphere should lead to a temperature increase by absorption of infrared radiation from the earth's surface. When these human factors are added to such other natural factors as volcanic eruptions, changes in solar activity, and resonances within the hydro-atmosphere, their effect can only be estimated in terms of direction, not of amount."	
1974	NSF: Science and the Challenges Ahead	"Human activity may be involved on an even broader scale in changing the global climate.... The cause of the cooling trend is not known with certainty. But there is increasing concern that man himself may be implicated, not only in the recent cooling trend but also in the warming temperatures over the last century.... By the middle of this century, the cooling effect of the dust particles more than compensated for the warming effect of the carbon dioxide, and world temperature began to fall."	National Science Board, National Science Foundation, "Science and the Challenges Ahead," 1974, https://archive.org/details/ sciencechallenge00nati.
1975	NAS: Understanding Climate Change: A Program for Action	"While the natural variations of climate have been larger than those that may have been induced by human activities during the past century, the rapidity with which human impacts threaten to grow in the future, and increasingly to disturb the natural course of events, is a matter of concern. These impacts include man's changes of the atmospheric composition and his direct interference with factors controlling the all important heat balance.... Of the two forms of pollution [CO_2 and aerosols], the carbon dioxide increase is probably the more influential at the present time in changing temperatures near the earth's surface. If both the CO_2 and particulate inputs to the atmosphere grow at equal rates in the future, the widely differing atmospheric residence times of the two pollutants means that the particulate effect will grow in importance relative to that of CO_2. There are other possible impacts of human activities that should be considered in projecting future climates."	National Academy of Sciences. *Understanding Climate Change: A Program for Action* (Washington, DC: National Academies Press, 1975), https://ia801806.us.archive.org/7/ items/understandingcli00unit/understanding cli00unit.pdf.
1983	World Climate Research Programme: Report of the Fifth Session of the Joint Scientific Committee	"The existing observational evidence does not allow concluding with confidence that the observed variations are caused specifically by the increasing concentration of atmospheric carbon dioxide. On the other hand, the surface warming indicated by the record of the global mean temperature is not inconsistent with the computed effect of the observed increase of atmospheric CO_2 since the beginning of the industrial revolution."	International Council of Scientific Unions, World Meteorological Organization, *Report of the Fifth Session of the Joint Scientific Committee (Hangzhou, China, 12-17 March 1984)*, September 1984, https://library. wmo.int/pmb_ged/wmo-td_1.pdf.

Year	Title	Key Quotation(s)	Reference
1985	Villach Conference Statement	"Based on analyses of observational data, the estimated increase in global mean temperature during the last one hundred years of between 0.3 and 0.7°C is consistent with the projected temperature increase attributable to the observed increase in CO_2 and other greenhouse gases, although it cannot be ascribed in a scientifically rigorous manner to these factors alone."	World Meteorological Organisation, *Report of the International Conference on the assessment of the role of carbon dioxide and of other greenhouse gases in climate variations and associated impacts,* Villach, Austria, October 9-15, 1985. WMO No. 661, 1986, https://web.archive.org/web/20131121040937/http://www.scopenvironment.org/downloadpubs/scope29/statement.html.
1987	Mankind's Impact on Climate: The Evolution of an Awareness	"There is now a strong consensus that the observed increase in the atmospheric concentrations of carbon dioxide and other infrared-absorbing trace gases is indeed warming the earth, and that this change is caused by mankind."	William W. Kellogg, "Mankind's Impact on Climate: The Evolution of an Awareness," *Climatic Change*, vol. 10 (1987), pp. 113-36, https://link.springer.com/content/pdf/10.1007%2FBF00140251.pdf.

Source: Compiled by CRS.

Table A-4. Excerpted Statements on the Human Contribution to Global Climate in Scientific Assessments of the NAS, the IPCC, and the USGCRP, 1977-2017

Year	Title	Key Quotation(s)	Reference
1977	NAS: Energy and Climate: Studies in Geophysics	"In the first half of the twentieth century ... a general warming of the earth occurred.... There is considerable evidence that, between the 1940s and about 1970, the climatic changes of the earlier part of this century had tended to undergo a reversal. Temperatures had mostly fallen." "Today, we find ourselves uneasy because through our use of energy we may be significantly disturbing the natural climate system. This uneasiness is justified."	National Academy of Sciences, *Energy and Climate* (Washington, DC: National Academies Press, 1977), https://www.nap.edu/download/12024.
1979	NAS: Carbon Dioxide and Climate: A Scientific Assessment	"None of the model calculations predicts negligible warming. The primary effect of an increase of CO_2 is to cause more absorption of thermal radiation from the earth's surface and thus to increase the air temperature in the troposphere." "We conclude that the predictions of CO_2-induced climate changes made with the various models examined are basically consistent and mutually supporting."	National Academy of Sciences, "Carbon Dioxide and Climate: A Scientific Assessment," Report of an Ad Hoc Study Group on Carbon Dioxide and Climate, Woods Hole, Massachusetts, July 23-27, 1979, to the Climate Research Board, Assembly of Mathematical and Physical Sciences, National Research Council, https://www.nap.edu/download/19856
1990	IPCC: Climate Change: The IPCC	"The size of the warming over the last century is broadly consistent with the predictions of climate models but is also of the same magnitude as natural climate	J. T. Houghton et al., *Climate Change. The IPCC Scientific Assessment.* Report prepared for

Table A-4. (Continued)

Year	Title	Key Quotation(s)	Reference
	Scientific Assessment - 1st Report	variability. If the sole cause of the observed warming were the human made greenhouse effect, then the implied climate sensitivity would be near the lower end of the range inferred from the models. The observed increase could be largely due to natural variability, alternatively this variability and other man-made factors could have offset a still larger man-made greenhouse warming. The unequivocal detection of the enhanced greenhouse effect from observations is not likely for a decade or more, when the commitment to future climate change will then be considerably larger than it is today."	Intergovernmental Panel on Climate Change by Working Group I (Cambridge, U.K.: Cambridge University Press, 1990), https://www.ipcc.ch/ ipccreports/far/wg_I/ ipcc_far_wg_I_full_report.pdf.
1994	NAS: Solar Influences on Global Change	"Taken collectively, the above evidence, although circumstantial, does suggest that solar variability could influence future global change.... Lack of knowledge of solar influences will limit the certainty with which anthropogenic climate change can be detected. But it is unlikely that solar influences on global change will be comparable to the expected anthropogenic influences."	National Academy of Sciences, *Solar Influences on Global Change* (Washington, DC: National Academies Press, 1994), https://www.nap.edu/download/4778.
1995	IPCC Second Assessment: Climate Change 1995	"Carbon dioxide remains the most important contributor to anthropogenic forcing of climate change; projections of future global mean temperature change and sea level rise confirm the potential for human activities to alter the Earth's climate to an extent unprecedented in human history.... [O]bservations suggest a 'discernible human influence on global climate.'" "Most of these studies have detected a significant change and show that the observed warming trend is unlikely to be entirely natural in origin."	B. Bolin et al., *Climate Change 1995: A report of the Intergovernmental Panel on Climate Change* (Cambridge, U.K.: Cambridge University Press, 1995), http://www.ipcc.ch/ipccreports/sar/wg_I/ ipcc_sar_wg_I_full_report.pdf.
2001	USGCRP: 1st National Climate Assessment	"The observed magnitude, pattern, and timing of the global warming indicate that the rising concentrations of CO_2 and other greenhouse gases caused by human activities are contributing significantly to the recent warming."	National Assessment Synthesis Team, *Climate Change Impacts on the United States: The Potential Consequences of Climate Variability and Change*, Report for the USGCRP (Cambridge, U.K.: Cambridge University Press, 2001), https://data.globalchange.gov/assets/e9/ 97/436129058f2107f4925aeec13ed8/nca-2000-foundation-report.pdf.
2001	NAS: Climate Change Science: An Analysis of Some Key Questions	"The changes observed over the last several decades are likely mostly due to human activities, but we cannot rule out that some significant part of these changes is also a reflection of natural variability."	National Academy of Sciences, *Climate Change Science: An Analysis of Some Key Questions* (Washington, DC: National Academy Press, 2001), https://www.nap.edu/download/10139.

Year	Title	Key Quotation(s)	Reference
2001	IPCC: Climate Change 2001: The IPCC Third Assessment Report	"There is new and stronger evidence that most of the warming observed over the last 50 years is attributable to human activities."	R. T. Watson et al. (eds.), *Climate Change 2001: Synthesis Report*, Contribution of Working Groups I, II, and III to the Third Assessment Report of the Intergovernmental Panel on Climate Change (Cambridge, U.K.: Cambridge University Press, 2001), https://www.ess.uci.edu/researchgrp/prather/files/2001IPCC_SyR-Watson.pdf.
2007	IPCC: The Fourth Assessment Report: Climate Change	"Most of the observed increase in global average temperatures since the mid-20th century is very likely [>90%] due to the observed increase in anthropogenic GHG concentrations."	R. K. Pachuri et al. (eds.), *Climate Change 2007: Synthesis Report*, Contribution of Working Groups I, II, and III to the Fourth Assessment Report of the Intergovernmental Panel on Climate Change (Geneva, Switzerland: IPCC Secretariat, 2007), http://www.slvwd.com/agendas/Full/2007/06-07-07/Item%2010b.pdf.
2009	USGCRP: 2nd National Climate Assessment	"The global warming of the past 50 years is due primarily to human-induced increases in heat trapping gasses.... Recent scientific assessments find that most of the warming of the Earth's surface over the past 50 years has been caused by human activities."	Thomas R. Karl et al. (eds.), *Global Climate Change Impacts in the United States* (Cambridge, UK: Cambridge University Press, 2009), https://downloads.globalchange.gov/usimpacts/pdfs/climate-impacts- report.pdf.
2010	NAS: Advancing the Science of Climate Change	"Climate change is occurring, is caused largely by human activities, and poses significant risks for—and in many cases is already affecting—a broad range of human and natural systems."	National Academy of Sciences, *Advancing the Science of Climate Change*, America's Climate Choices: Panel on Advancing the Science of Climate Change (Washington, DC: National Academy Press, 2010), https://www.nap.edu/download/12782.
2012	NAS: Climate Change: Evidence, Impacts, and Choices	"The overwhelming majority of climate scientists agree that human activities, especially the burning of fossil fuels (coal, oil, and gas), are responsible for most of the climate change currently being observed."	National Academy of Sciences, *Climate Change: Evidence, Impacts, and Choices*, 2012, http://nas-sites.org/americasclimatechoices/files/2012/06/19014_cvtx_R1.pdf
2014	NAS: Climate Change: Evidence and Causes	"Rigorous analysis of all data and lines of evidence shows that most of the observed global warming over the past 50 years or so cannot be explained by natural causes and instead requires a significant role for the influence of human activities."	National Academy of Sciences, *Climate Change: Evidence and Causes*, 2014, http://dels.nas.edu/resources/static- assets/exec-office-other/climate-change- full.pdf
2014	IPCC: The Fifth Assessment Report: Climate Change 2014	"[Anthropogenic GHG emissions] effects, together with those of other anthropogenic drivers, have been detected throughout the climate system and are extremely likely [>95%] to have been the dominant cause of the observed warming since the mid-20th century."	R. K. Pachuri et al. (eds.), *Climate Change 2014: Synthesis Report*, Contribution of Working Groups I, II, and III to the Fifth Assessment Report of the Intergovernmental Panel on Climate Change (Geneva, Switzerland: IPCC Secretariat, 2014),

Table A-4. (Continued)

Year	Title	Key Quotation(s)	Reference
			http://www.ipcc.ch/ pdf/assessment-report/ar5/syr/ SYR_AR5_FINAL_full_wcover.pdf.
2014	USGCRP: 3rd National Climate Assessment	"The majority of the warming at the global scale over the past 50 years can only be explained by the effects of human influences, especially the emissions from burning fossil fuels (coal, oil, and natural gas) and from deforestation."	Jerry M. Melillo et al., (eds.), *Highlights of Climate Change Impacts in the United States: The Third National Climate Assessment*, USGCRP (Washington, DC: U.S. Government Printing Office, 2014), http://www.globalchange.gov/sites/ globalchange/files/NCA3_Highlights_ LowRes-small-FINAL_posting.pdf.
2017	USGCRP: Climate Science Special Report: Fourth National Climate Assessment	"It is extremely likely [>95%] that human influence has been the dominant cause of the observed warming since the mid-20th century. For the warming over the last century, there is no convincing alternative explanation supported by the extent of the observational evidence."	Donald J. Weubbles et al., *Climate Science Special Report: Fourth National Climate Assessment, Volume I*, USGCRP, 2017, https://science2017.globalchange.gov/ downloads/CSSR2017_FullReport.pdf.

Source: Compiled by CRS.

Notes: The documents cited reflect the evolution of select aspects of climate change science. This list is not comprehensive of, and is not intended to highlight, all aspects of the scientific efforts in climate research.

In: Climate Change
Editor: Jeremy Nesbitt

ISBN: 978-1-53617-216-4

Chapter 2

CLIMATE CHANGE: ACTIVITIES OF SELECTED AGENCIES TO ADDRESS POTENTIAL IMPACT ON GLOBAL MIGRATION*

United States Government Accountability Office

ABBREVIATIONS

ACP	African, Caribbean, and Pacific States
ADB	Asian Development Bank
CRS	Congressional Research Service
DHS	Department of Homeland Security
DOD	Department of Defense
GDP	Gross Domestic Product
IOM	International Organization for Migration
IPCC	Intergovernmental Panel on Climate Change

* This is an edited, reformatted and augmented version of the United States Government Accountability Office Report to Congressional Requesters, Publication No. GAO-19-166, January 2019.

LDCF	Least Developed Countries Fund
PIER	Private Investment for Enhanced Resilience
State	Department of State
State/OES	State's Bureau of Oceans and International Environmental and Scientific Affairs
State/PRM	State's Bureau of Population, Refugees, and Migration
TPS	Temporary Protected Status
UN	United Nations
UNEP	United Nations Environment Programme
UNHCR	United Nations High Commissioner for Refugees
USAID	U.S. Agency for International Development
USAID/FFP	USAID's Office of Food for Peace
USAID/OFDA	USAID's Office of U.S. Foreign Disaster Assistance
USIP	United States Institute of Peace

WHY GAO DID THIS STUDY

The effects of climate change, combined with other factors, may alter human migration trends across the globe, according to the International Organization for Migration. For example, climate change can increase the frequency and intensity of natural disasters, causing populations to move from an area. Climate change can also intensify slow-onset disasters, such as drought, crop failure, or sea level rise, potentially altering longer-term migration trends.

GAO was asked to review how U.S. agencies address climate change as a potential driver of global migration. For State, USAID, and DOD, this chapter (1) describes executive branch actions related to climate change and migration from fiscal years 2014 through 2018; (2) examines the extent to which the agencies discussed the potential effects of climate change on migration in their plans and risk assessments; and (3) describes agency activities on the issue. GAO analyzed documents on administration priorities; reviewed agency plans, risk assessments, and documentation of agency activities; and interviewed agency officials.

WHAT GAO RECOMMENDS

GAO recommends that State provide missions with guidance that clearly documents its process for climate change risk assessments for country strategies. In commenting on a draft of this chapter, State indicated that it would update its integrated country strategy guidance and will specifically note that missions have the option to provide additional information on climate resilience and related topics.

WHAT GAO FOUND

From fiscal years 2014 through 2018, a variety of executive branch actions related to climate change—such as executive orders and strategies—affected the Department of State (State), the U.S. Agency for International Development (USAID), and the Department of Defense (DOD), including their activities that could potentially address the nexus of climate change and migration. For example, a fiscal year 2016 presidential memorandum—rescinded in 2017— required agencies to develop implementation plans to identify the potential impact of climate change on human mobility, among other things. In general, however, climate change as a driver of migration was not a focus of the executive branch actions. For example, a fiscal year 2014 executive order—also rescinded in 2017—requiring agencies to prepare for the impacts of climate change did not highlight migration as a particular concern.

State, USAID, and DOD have discussed the potential effects of climate change on migration in agency plans and risk assessments. For example, State and USAID required climate change risk assessments when developing country and regional strategies, and a few of the strategies reviewed by GAO identified the nexus of climate change and migration as a risk. However, State changed its approach in 2017, no longer providing missions with guidance on whether and how to include climate change risks in their integrated country strategies. In doing so, State did not include in its 2018 guidance to the missions any information on how to include climate change risks, should the missions choose to do so. Without clear guidance, State may miss opportunities to identify and address issues related to climate change as a potential driver of migration.

The three agencies have been involved in climate change related activities but none were specifically focused on the nexus with global migration. For example, USAID officials said that the agency's adaptation efforts, such as its Pastoralist Areas Resilience Improvement through Market Expansion project in Ethiopia, were the most likely to include activities, such as enhancing resilience, that can indirectly address the issue of climate change as a driver of migration.

January 17, 2019
Congressional Requesters

Around the world, climate change is predicted to affect precipitation levels, increase temperatures, and contribute to more frequent natural disasters like flooding and drought, among other effects, according to the Intergovernmental Panel on Climate Change (IPCC).[1] At the same time, recent trends suggest that the number of global migrants will continue to increase,[2] according to the International Organization for Migration (IOM).[3] While many factors, including lack of economic opportunity and political instability, influence the decision to migrate, these international organizations predict that climate change will further increase global human migration, but they do not know to what extent. Nonetheless, the

---

[1] Intergovernmental Panel on Climate Change, *Climate Change 2014: Synthesis Report* (Geneva, Switzerland: 2014).

[2] In this chapter we use "migration" to refer to global, human migration. In keeping with the International Organization for Migration (IOM), we define migration as including both the movement of people within a country, and across international borders, either temporarily or permanently.

[3] International Organization for Migration, *World Migration Report 2018* (Geneva, Switzerland: 2017).

potential link between climate change and migration has raised both humanitarian and national security concerns for the U.S. government.

As a result of the risks that climate change poses to environmental and economic systems,[4] in February 2013 we placed *Limiting the Federal Government's Fiscal Exposure by Better Managing Climate Change Risks* on our High Risk List.[5] As we reported in February 2017, for example, from fiscal years 2005 through 2014, the federal government obligated at least $277.6 billion across 17 federal departments and agencies for disaster assistance programs and activities.[6] Extreme weather events—such as floods, drought, and hurricanes—are expected to increase or worsen because of climate change, according to the U.S. Global Change Research Program.[7] In February 2017, we reported that the federal government needs a cohesive strategic approach with strong leadership and the authority to manage climate change risks that encompasses the entire range of related federal activities and addresses all key elements of strategic planning.[8]

You asked us to review issues related to human migration due to climate change. In this chapter, we (1) describe executive branch actions related to climate change and migration from fiscal years 2014 through 2018; (2) examine the extent to which the Department of State (State), the U.S. Agency for International Development (USAID), and the Department of Defense (DOD) have discussed the potential effects of climate change on migration in their plans and risk assessments; and (3) describe State,

---

[4] GAO, *High-Risk Series: An Update*, GAO-15-290 (Washington, D.C.: Feb. 11, 2015). GAO's high-risk program identifies government operations with greater vulnerabilities to fraud, waste, abuse, and mismanagement or the need for transformation to address economy, efficiency, or effectiveness challenges.

[5] GAO, *High-Risk Series: An Update*, GAO-13-283 (Washington, D.C.: February 2013). See also GAO, *High-Risk Series: Progress on Many High-Risk Areas, While Substantial Efforts Needed on Others*, GAO-17-317 (Washington, D.C.: Feb. 15, 2017).

[6] GAO-17-317.

[7] Jay, A., D.R. Reidmiller, C.W. Avery, D. Barrie, B.J. DeAngelo, A. Dave, M. Dzaugis, M. Kolian, K.L.M. Lewis, K. Reeves, and D. Winner, 2018: Overview. In *Impacts, Risks, and Adaptation in the United States: Fourth National Climate Assessment, Volume II* [Reidmiller, D.R., C.W. Avery, D.R. Easterling, K.E. Kunkel, K.L.M. Lewis, T.K. Maycock, and B.C. Stewart (eds.)]. U.S. Global Change Research Program, Washington, DC, USA.

[8] GAO-17-317.

USAID, and DOD activities, if any, that are related to climate change and global migration.

To describe executive branch actions related to climate change and migration from fiscal years 2014 through 2018, we reviewed key documents related to climate change that were developed during this time period. These documents included executive orders, budget requests, and strategies. We chose fiscal years 2014 through 2018 as our time frame based on our review of recent executive orders related to climate change. To examine State, USAID, and DOD discussions of the potential effects of climate change on migration in their plans and risk assessments, we reviewed relevant documents completed since fiscal year 2014. These documents included agency adaptation plans and country and regional strategies. We selected State, USAID, and DOD because the agencies' missions of diplomacy, development, and defense provide the foundation for promoting and protecting U.S. interests abroad.[9] To describe State, USAID, and DOD activities that are related to climate change and global migration, we identified and examined agency activities potentially related to the issue. For example, we analyzed State and USAID data such as project descriptions for activities that received adaptation funding from the Global Climate Change Initiative. We determined that the USAID and State adaptation project data were sufficiently reliable for the purposes of describing their efforts. For all three objectives we interviewed officials from State, USAID, and DOD. Further details on our scope and methodology can be found in appendix I.

We conducted this performance audit from October 2017 to January 2019 in accordance with generally accepted government auditing standards. Those standards require that we plan and perform the audit to obtain sufficient, appropriate evidence to provide a reasonable basis for our findings and conclusions based on our audit objectives. We believe that the

---

[9] We also reviewed documents and met with officials from the Departments of Homeland Security and Treasury, the National Oceanic and Atmospheric Administration, the U.S. Environmental Protection Agency, and the U.S. Global Change Research Program. We did not include these agencies in our review because they did not have assessments or activities related to climate change as a driver of global migration during the time frame of our review.

evidence obtained provides a reasonable basis for our findings and conclusions based on our audit objectives.

BACKGROUND

Potential Impacts of Climate Change on Migration

According to international and U.S. government sources, climate change poses serious risks to many of the physical and ecological systems upon which society depends, although the exact details of these impacts are uncertain.[10] Climate change may intensify slow-onset disasters, such as drought, crop failure, and sea level rise. Climate change is also increasingthe frequency and intensity of extreme weather events, including sudden-onset disasters, such as floods, according to key scientific assessments.[11] These effects of climate change may alter existing migration trends across the globe, according to IOM. (See appendix II for further discussion of climate change as a driver of migration in seven geographic regions.) For example, sea level rise, a slow-onset disaster, may result in the salinization of soil and drinking water, thereby undermining a country

---

[10] Intergovernmental Panel on Climate Change, *Climate Change 2014: Synthesis Report* (Geneva, Switzerland: 2014); International Organization for Migration, *The Atlas of Environmental Migration* (New York: Routledge, 2017); Kumari Rigaud, Kanta, Alex de Sherbinin, Bryan Jones, Jonas Bergmann, Viviane Clement, Kayly Ober, Jacob Schewe, Susana Adamo, Brent McCusker, Silke Heuser, and Amelia Midgley. 2018. *Groundswell: Preparing for Internal Climate Migration*. Washington, DC: The World Bank; U.S. Department of Defense, *National Security Implications of Climate-Related Risks and a Changing Climate, 8-6475571* (July 2015); and U.S. Agency for International Development, *Climate Change and Conflict in Africa and Latin America: Findings and Preliminary lessons from Uganda, Ethiopia, and Peru* (2013).

[11] Jerry M. Melillo, Terese (T.C.) Richmond, and Gary W. Yohe, eds., Climate Change Impacts in the United States: The Third National Climate Assessment, (Washington, D.C.: U.S. Global Change Research Program, May 2014) and Intergovernmental Panel on Climate Change, 2014: Climate Change 2014: Impacts, Adaptation, and Vulnerability. Part A Global and Sectoral Aspects. Contribution of Working Group II to the Fifth Assessment Report of the Intergovernmental Panel on Climate Change [Field, C.B., V.R. Barros, D.J. Dokken, K.J. Mach, M.D. Mastrandrea, T.E. Bilir, M. Chatterjee, K.L. Ebi, Y.O. Estrada, R.C. Genova, B. Girma, E.S. Kissel, A.N. Levy, S. MacCracken, P.R. Mastrandrea, and L.L. White (eds.)]. Cambridge University Press, Cambridge, United Kingdom and New York, NY, USA, 1132 pp.

or community's ability to sustain livelihoods and maintain critical services, which could cause some people to migrate. Sudden-onset disasters may also contribute to migration as people flee natural disasters, in most cases leading to temporary displacement. For example, people may either voluntarily migrate, or be forced to migrate, to earn money needed to rebuild damaged homes after flooding, especially as extreme weather events increase in intensity and number. If unable or unwilling to migrate, people may find themselves trapped or choosing to stay in deteriorating conditions. Sources agree that the effects of climate change generally impact internal migration, while migration across international borders due to climate change is less common.[12]

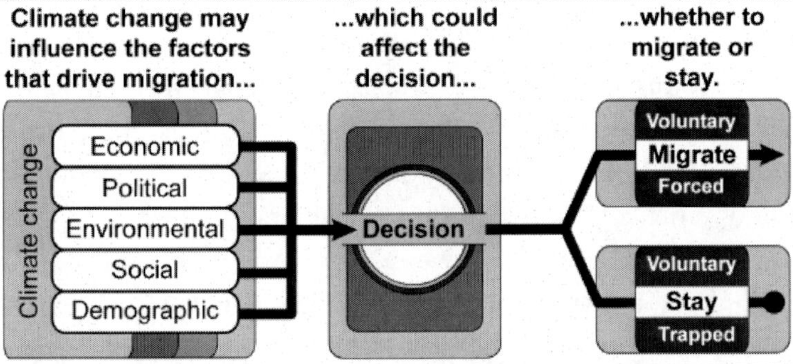

Source: GAO analysis of United Kingdom government Office for science information. | GAO-19-166.

Note: Additional factors may influence migration. These include the cost of moving, social networks, and diasporic links, among others. Further, personal or household characteristics, such as age, religion, or education, may make migration more or less likely.

Figure 1. Climate Change as a Potential Influence on Other Factors that Drive Migration.

[12] Schuyler Null and Lauren Herzer Risi, Woodrow Wilson International Center for Scholars, USAID Office of Conflict Management and Mitigation Discussion Paper, Navigating Complexity: Climate, Migration, and Conflict in a Changing World (November 2016); Kumari Rigaud, Kanta, Alex de Sherbinin, Bryan Jones, Jonas Bergmann, Viviane Clement, Kayly Ober, Jacob Schewe, Susana Adamo, Brent McCusker, Silke Heuser, and Amelia Midgley. 2018. Groundswell: Preparing for Internal Climate Migration. Washington, DC: The World Bank; The United Kingdom Government Office for Science, Foresight: Migration and Global Environmental Change: Final Project Report (2011).

In deciding whether to migrate, people weigh multiple factors including economic and political factors, social or personal motives, or demographic pressures. The effects of climate change add another layer of complexity to this decision, but there is debate about the role climate change plays in migration. Figure 1 depicts how climate change may influence other factors that drive the decision to migrate or stay.

There are limitations to reliably estimating the number of people displaced by climate change because there are no reliable global estimates for those migrating due to slow-onset disasters, and estimates for those migrating due to sudden-onset disasters are based on limited data, according to IOM. The lack of reliable data is due in part to the multi-causal nature of migration. Further, IOM notes that forecasts for the number of environmental migrants by 2050 vary from 25 million to 1 billion.[13] They and others have questioned the methodologies used to arrive at even these broad estimates.

Climate Change Impacts on Migration that May Affect National Security

Migration, potentially driven by climate change, may contribute to instability and result in national security challenges, according to some international organizations and national governments.[14] For example, an influx of migrants to a city may put pressure on existing resources, resulting in tensions between new migrants and residents, or between the population and its government. The U.S. Global Change Research Program has also stated that migration, such as displacement resulting from extreme

[13] International Organization for Migration, *IOM Outlook on Migration, Environment and Climate Change* (Geneva, 2014).

[14] UN General Assembly, Climate Change and Its Possible Security Implication: Report to the Secretary General (2009); Intergovernmental Panel on Climate Change, Climate Change 2014: Impacts, Adaptation, and Vulnerability (Cambridge, United Kingdom: Cambridge University Press, 2017); The United Kingdom Government Office for Science, Foresight: Migration and Global Environmental Change: Final Project Report (2011); Guy S. Goodwin-Gill and Jane McAdam, Climate Change, Disasters, and Displacement (UNHCR–The UN Refugee Agency, 2017).

weather events, is a potential national security issue.[15] At different times, the United Nations General Assembly and, in 2014, DOD have deemed climate change to be a threat multiplier, as the effects of climate change could increase competition for resources, reduce government capacity, and threaten livelihoods, thereby causing instability and migration.[16] Further, the U.S. intelligence community considers climate change to increase the risks of humanitarian disasters, conflict, and migration.[17]

Identifying the cause of a conflict, however, is complicated, and experts debate the connections linking climate, migration, and national security. For example, IOM has reported that existing evidence on climate migration and instability must be considered with caution.[18] Further, some studies stress that other factors can mitigate the effects of climate change on migration and stability, including governance and community resilience, as the World Bank has reported.[19]

U.S. Government Agency Roles Related to Climate Change

State, USAID, and DOD are among the U.S. government agencies with a role in responding to issues related to climate change, including as a driver of migration.

[15] Smith, J.B., M. Muth, A. Alpert, J.L. Buizer, J. Cook, A. Dave, J. Furlow, K. Preston, P. Schultz, and L. Vaughan, 2018: Climate Effects on U.S. International Interests. In Impacts, Risks, and Adaptation in the United States: Fourth National Climate Assessment, Volume II [Reidmiller, D.R., C.W. Avery, D.R. Easterling, K.E. Kunkel, K.L.M. Lewis, T.K. Maycock, and B.C. Stewart (eds.)]. U.S. Global Change Research Program, Washington, DC, USA.

[16] UN General Assembly, Climate Change and Its Possible Security Implication: Report to the Secretary General (2009); and United States Department of Defense, 2014 Quadrennial Defense Review (Washington, D.C.: 2014).

[17] United States Office of the Director of National Intelligence, Worldwide Threat Assessment of the U.S. Intelligence Community (February, 2018).

[18] International Organization for Migration, The Atlas of Environmental Migration (New York: Routledge, 2017).

[19] Kumari Rigaud, Kanta, Alex de Sherbinin, Bryan Jones, Jonas Bergmann, Viviane Clement, Kayly Ober, Jacob Schewe, Susana Adamo, Brent McCusker, Silke Heuser, and Amelia Midgley. 2018. *Groundswell: Preparing for Internal Climate Migration*. Washington, DC: The World Bank.

- State interacts with foreign governments and international organizations focused on climate change and migration primarily through the Bureau of Occans and International Environmental and Scientific Affairs (State/OES) and the Bureau of Population, Refugees, and Migration (State/PRM).
- USAID supports a range of development programs that help to mitigate the effects of climate change through the Bureaus for Economic Growth, Education and Environment; Democracy, Conflict and Humanitarian Assistance; Food Security; Asia; and Africa; and individual USAID missions.[20]_Additionally, USAID's Offices of U.S. Foreign Disaster Assistance (USAID/OFDA) and Food for Peace (USAID/FFP) lead and coordinate the U.S. government's emergency responses to sudden- and slow-onset disasters, and complex emergencies overseas.[21]
- DOD assists in the United States' humanitarian response to sudden-onset disasters abroad through its six geographic combatant commands,[22] with support from the Assistant Secretary of Defense for Special Operations and Low Intensity Conflict and the Joint Staff's Office of Humanitarian Engagement.[23]

---

[20] USAID officials said that, pending congressional approval, the technical leadership of USAID's climate adaptation work will move to the proposed Bureau for Resilience and Food Security as part of the Center for Resilience, and other climate technical leadership functions will move to the proposed Bureau for Development, Democracy and Innovation.

[21] USAID/OFDA is the lead federal coordinator for international disaster assistance. USAID/FFP is the U.S. government lead in providing emergency global food assistance. USAID said that, pending congressional approval, it intends to merge and restructure OFDA and Food for Peace to form the Bureau for Humanitarian Assistance in order to consolidate its response to natural disasters, famines, and man-made crises.

[22] To perform its variety of missions around the world, DOD operates six geographic combatant commands which manage all military operations in designated areas of responsibility: U.S. Africa Command, U.S. Central Command, U.S. European Command, U.S. Northern Command, U.S. Indo-Pacific Command, and U.S. Southern Command.

[23] The Joint Staff is responsible for assisting the Chairman of the Joint Chiefs of Staff, the military advisor to the President, in accomplishing his responsibilities for the unified strategic direction of the combatant forces; their operation under unified command; and their integration into a team of land, naval, and air forces. The Joint Staff is tasked to provide advice and support to the Chairman and the Joint Chiefs on matters including personnel, intelligence doctrine and architecture, operations and plans, logistics, strategy, policy, communications, cyberspace, joint training and education, and program evaluation.

EXECUTIVE BRANCH ACTIONS RELATED TO CLIMATE CHANGE AND MIGRATION FROM FISCAL YEARS 2014 THROUGH 2018

Climate change as a driver of migration was not a focus of the policy documents we reviewed for either the current or previous administrationsduring fiscal years 2014 through 2018. Our review of executive actions, budget requests, and executive branch strategies that affected State, USAID, and DOD found only brief mentions of climate change as a driver of migration. None of the documents we reviewed reflected a priority for assessing or addressing climate change as a driver of migration, although these documents reflect a shift in administrations' climate change priorities more generally.

Executive Actions

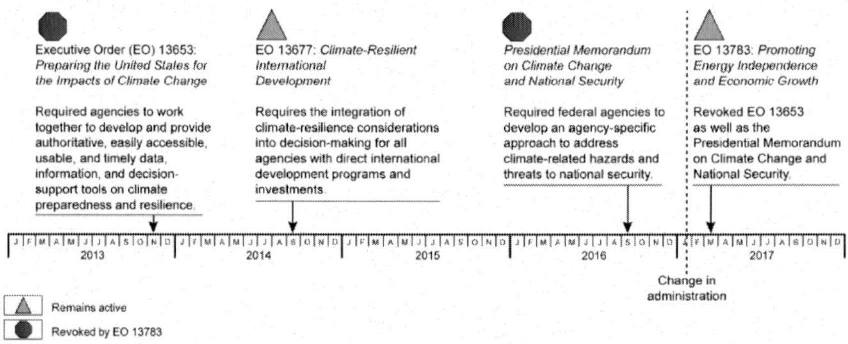

Source: GAO analysis of executive actions. | GAO-19-166.

Note: The descriptions listed are examples of climate change-related requirements and are not the only requirements contained in these actions.

Figure 2. Timeline of Recent Executive Actions Related to Climate Change.

The previous administration issued two executive orders and a presidential memorandum related to climate change. These executive actions had a policy of improving climate preparedness and resilience, factoring climate-resilience considerations into agencies' international development decisions, and creating forums for interagency coordination. In March 2017, the current administration issued a subsequent executive order revoking some of the previous executive actions related to climate change. See figure 2 for a timeline of these executive actions.

The previous administration issued three executive actions related to climate change, which included requirements focused on agencies' considerations of the impacts of climate change and established forums for interagency coordination. The current administration issued an executive action related to energy independence and climate change.

- *Executive Order 13653: Preparing the United States for the Impacts of Climate Change.* Executive Order 13653 stated that agencies—including State, USAID, and DOD—shall, among other things, develop, implement, and update comprehensive Agency Adaptation Plans that integrate consideration of climate change into agency operations and overall mission objectives.[24] Executive Order 13653 also established the Council on Climate Preparedness and Resilience.
- *Executive Order 13677: Climate-Resilient International Development.* Executive Order 13677 requires State, USAID, and other U.S. government agencies with direct international development programs and investments to incorporate climate-resilience considerations into decision making by assessing climate-related risks to agency strategies, and to adjust relevant strategies as appropriate, among other things.[25] Executive Order 13677 also established the Working Group on Climate-Resilient International Development as part of the Council on Climate Preparedness and Resilience.

---

[24] Exec. Order No. 13,653, Preparing the United States for the Impacts of Climate Change, 78 Fed. Reg. 66,819 (Nov. 1, 2013).

[25] Exec. Order No. 13,677, *Climate-Resilient International Development*, 79 Fed. Reg. 58,231 (Sept. 23, 2014). The executive order did not identify DOD as an agency with direct international development programs and investments.

- *2016 Presidential Memorandum on Climate Change and National* Security. The 2016 presidential memorandum required, among other things, that agencies, including State, USAID, and DOD, develop an agency-specific approach to address climate-related threats to national security. It also required agencies to develop implementation plans that would describe how they would identify the potential impact of climate change on human mobility, including migration and displacement, and the resulting impacts on national security, among other requirements, and stated that the effects of climate change can lead to population migration within and across international borders, spur crises, and amplify or accelerate conflict in countries or regions already facing instability.[26] The 2016 memorandum also established the Climate and National Security Working Group.
- *Executive Order 13783, Promoting Energy Independence and Economic Growth.* Executive Order 13783 revoked Executive Order 13653 and the 2016 presidential memorandum, among other things, as seen in figure 2.[27]

Presidential Budget Requests for Fiscal Years 2017 and 2018

Priorities related to climate change shifted between the past two administrations as reflected in a recent budget request that reduced some climate change funding affecting U.S. foreign assistance.

- 2017 Presidential Budget Request. The previous administration stated in its fiscal year 2017 budget request that "the challenge of climate change will define the contours of this century more dramatically than any other" and that "it is imperative for the United States to couple action on climate change at home with

[26] Presidential Memorandum on Climate Change and National Security (Sept. 21, 2016).
[27] Exec. Order No. 13,783, *Promoting Energy Independence and Economic Growth*, 82 Fed. Reg. 16,093 (Mar. 28, 2017).

leadership internationally."[28] The fiscal year 2017 budget request sought $1.3 billion in discretionary funding to advance the goals of the Global Climate Change Initiative, which was established in 2010 and aimed to promote resilient, low-emission development, and integrate climate change considerations into U.S. foreign assistance.[29] The $1.3 billion in requested funding included $750 million in U.S. funding for the Green Climate Fund, a multilateral trust fund designed to foster resilient low-emission development in developing countries.[30]

- 2018 Presidential Budget Request. The current administration, in its fiscal year 2018 budget request, did not include any funding for the Global Climate Change Initiative.[31] In addition, the current administration's budget request stated that it "Eliminate[d] the Global Climate Change Initiative and fulfill[ed] the President's pledge to cease payments to United Nations' (UN) climate change programs by eliminating U.S. funding related to the Green Climate Fund. . ."[32]

---

[28] Office of Management and Budget, Budget of the U.S. Government, Fiscal Year 2017 (Feb. 9, 2016).

[29] The Global Climate Change Initiative was divided into three main programmatic initiatives: (1) Adaptation assistance, (2) Clean Energy assistance, and (3) Sustainable Landscapes assistance. The Global Climate Change Initiative has been funded through programs at the Departments of State and Treasury, and USAID. Funds for these programs have been requested in the President's budget under the International Affairs Function 150 account for State, Foreign Operations, and Related Programs, according to the Congressional Research Service (CRS). Many Global Climate Change Initiative activities have been funded at agency sub-account levels, with allocations left to the discretion of the agencies, under congressional consultation, reported CRS. Recent budget authority for the Global Climate Change Initiative, according to CRS, was approximately $945 million in fiscal year 2010, $819 million in fiscal year 2011, $858 million in fiscal year 2012, $841 million in fiscal year 2013, $834 million in fiscal year 2014, and $824 million in fiscal year 2015. Department of Treasury officials told us that they have not done any specific work—assessments, policy papers, or programming—on migration due to climate change; thus we did not include Treasury in our review.

[30] The United States made a multi -year pledge of $3 billion to the initial resource mobilization of the Green Climate Fund; the first U.S. contribution of $500 million to the Green Climate Fund was requested in the President's fiscal year 2016 budget request.

[31] Office of Management and Budget, Budget of the U.S. Government: A New Foundation for American Greatness, Fiscal Year 2018 (May 23, 2017).

[32] Office of Management and Budget, *America First: A Budget Blueprint to Make America Great Again* (March 16, 2017). While the Consolidated Appropriations Act, 2018, did not include funding for the Green Climate Fund, it did provide no less than $123.5 million for

Strategy Documents Affecting State, USAID, and DOD

Some strategies from the current and previous administrations that affect State, USAID, and DOD, among other agencies, reflect a shift in priorities related to climate change. For example, the previous administration cited climate change as a "top strategic risk" in its 2015 National Security Strategy and stated that climate change is an urgent and growing threat to U.S. national security, contributing to increased natural disasters, refugee flows, and conflicts over basic resources like food and water.[33] The current administration does not discuss climate change in its 2017 National Security Strategy.[34] Additionally, State and USAID have a Joint Strategic Plan to help the agencies achieve the objectives of the National Security Strategy. The previous State-USAID Joint Strategic Plan included a strategic goal on "promoting the transition to a low-emission, climate-resilient world" that proposed leading international actions to combat climate change.[35] The current State-USAID Joint Strategic Plan does not have a climate change goal.[36]

State, USAID, and DOD Have Discussed the Potential Effects of Climate Change on Global Migration, but State Does Not Provide Clear Risk Assessment Guidance

State, USAID, and DOD were required by executive orders to assess climate change-related risks to their missions and, for State and USAID, to their strategies, among other things. In response to Executive Order 13653,

---

Sustainable Landscapes programs, one of the initiatives previously under the Global Climate Change Initiative. See Pub. L. No. 115 -141 (Mar. 23, 2018). In addition, according to State's Bureau of International Organization Affairs, State is no longer providing funding for the United Nations Framework Convention on Climate Change and Intergovernmental Panel on Climate Change.

[33] The White House, *National Security Strategy* (February 2015).

[34] .The White House, *National Security Strategy* (December 2017).

[35] State-USAID Joint Strategic Plan FY 2014-2017 (March 2014).

[36] State-USAID Joint Strategic Plan FY 2018-2022 (February 2018).

which has since been revoked, the agencies completed adaptation plans that integrated considerations of climate change into agency operations and overall mission objectives. In response to Executive Order 13677, which has not been revoked, State and USAID developed processes for climate change risk assessments for their country and regional planning documents.[37] Although these executive orders did not require a specific assessment of climate change as a driver of migration, all three agencies have discussed the effects of climate change on migration in their adaptation plans and risk assessments. However, State lacks clear guidance on its process for assessing climate change-related risks to its integrated country strategies.

Agencies Discussed the Effects of Climate Change on Migration in Their 2014 Adaptation Plans

State, USAID, and DOD each completed adaptation plans in 2014 that included limited discussions of migration as one potential effect of climate change. Executive Order 13653 directed the agencies to develop or continue to develop, implement, and update comprehensive Agency Adaptation Plans that integrate consideration of climate change into agency operations and overall mission objectives. Each adaptation plan was to include, among other things, a description of how the agency would consider the need to improve climate adaptation and resilience.

- State. In its 2014 adaptation plan, State included a brief discussion of climate change as one of multiple factors that potentially will drive migration and impact its mission. State reported that the specific impacts of climate change on the ability of the department to promote peace and stability in regions of vital interest to the

[37] Exec. Order No. 13,677, Climate-Resilient International Development, 79 Fed. Reg. 58,231 (Sept. 23, 2014). Executive Order 13677 required State, USAID, and other U.S. government agencies with direct international development programs and investments to incorporate climate-resilience considerations into decision-making by assessing climate-related risks to and vulnerabilities in existing strategies, and adjusting relevant strategies, as appropriate. The executive order did not list DOD as a relevant agency.

United States were unknown. For example, according to the plan, an increase in heavy precipitation events around the world could damage the electric grid and transportation and energy water infrastructure, upon which State depends, making it difficult to maintain operations and diplomatic relations. In its plan, State reported that climate change impacts may threaten international peace, civil stability, and economic growth through aggravating existing problems related to poverty and environmental degradation. Further, environmental and poverty- related issues and regional instability could stress relationships with some foreign governments. However, the plan noted that specific impacts of climate change on conflict, migration, terrorism, and complex disasters were still unknown.

- USAID. In its 2014 adaptation plan, USAID included a brief discussion of migration as one potential effect of climate change that could also impact security.[38] USAID stated that the impact of climate change on its programs and operations, if left unaddressed, could compromise the agency's ability to achieve its mission. Further, USAID's plan referred to increased migration as a potential risk of climate change. Flooding and other extreme climate events can result in increased migration, among other impacts, that could affect existing and planned USAID programming. In particular, programs in areas like agriculture and food security, global health, water and sanitation, infrastructure, and disaster readiness and humanitarian response are vulnerable to climate change, according to USAID. In the infrastructure area, climate change may necessitate new protective measures for coastal homes and infrastructure, and in some cases even mass evacuations or permanent migration. USAID stated that climate change could further reduce or alter the distribution of already limited resources like food and water, or force temporary or permanent migration of communities. According to the plan, in

[38] USAID rescinded its June 2014 Climate Change Adaptation Plan on May 11, 2017.

areas with high risk factors for conflict, climate change stresses can aggravate tensions and contribute to conflict.

- DOD. In its 2014 adaptation roadmap, DOD included a brief discussion of migration as one of multiple potential effects of climate change that could impact national security. DOD referred to climate change as a threat multiplier that can aggravate other risks around the world, with migration being one effect that could increase requests for DOD to provide assistance.[39] The roadmap stated that as climate change affects the availability of food and water, human migration, and competition for natural resources, the department's unique capability to provide logistical, material, and security assistance on a massive scale or in rapid fashion may be called upon with increasing frequency. Furthermore, DOD stated that the impacts of climate change may cause instability in other countries by, among other things, impairing access to food and water, damaging infrastructure, uprooting and displacing large numbers of people, and compelling mass migration. These developments, according to the department, could undermine already fragile governments that are unable to respond effectively, or challenge currently stable governments, as well as increase competition and tension between countries vying for limited resources.

Few of the State and USAID Risk Assessments We Reviewed Identified the Nexus of Climate Change and Migration as a Risk

In response to Executive Order 13677, State and USAID developed processes for climate change risk assessments for their country and regional planning documents.[40] Though these assessments are not specific

---

[39] DOD developed the 2014 *Roadmap* to fulfill requirements of a Climate Change Adaptation Plan found in Executive Orders 13514 on *Federal Leadership in Environmental, Energy, and Economic Performance* and 13653 on *Preparing the U.S. for the Impacts of Climate Change.*

[40] Exec. Order No. 13,677, Climate-Resilient International Development, 79 Fed. Reg. 58,231 (Sept. 23, 2014). Executive Order 13677 required State, USAID, and other U.S. government

to migration, a few of the assessments identified the nexus of climate change and migration.

- State. State required climate change risk assessments for all new integrated country strategies drafted in 2016 or later.[41] We reviewed 10 integrated country strategies from the two regions that were the first to implement the climate change risk assessment requirement— Africa, and East Asia and the Pacific. All 10 of the strategies included climate change risk assessments, one of which—Cambodia—identified migration as a risk for the country.[42] The Cambodia strategy states that internal migration due to climate change hinders access to health care and the prevention of infectious diseases like malaria. We also reviewed 10 strategies from State's functional and regional bureaus for assessments of climate-related risks, including 3 functional bureau strategies (State/PRM, State/OES, and State's Bureau of International Organization Affairs) and 7 regional bureau strategies. All of the functional bureau strategies we reviewed identified climate change as a risk and State/PRM cited the impact of climate change on

agencies with direct international development programs and investments to incorporate climate-resilience considerations into decision-making by assessing climate- related risks to and vulnerabilities in existing strategies, and adjusting relevant strategies, as appropriate. The executive order did not list DOD as a relevant agency.

[41] State's 2016 integrated country strategy guidance stated that all missions should perform a climate change risk assessment, and if one or more goals and objectives are determined medium-high risk, missions must document the climate-risk screening process in an annex. Those only identifying low risks may include a summary statement explaining how the mission came to this determination. An integrated country strategy is a multi-year plan that articulates the U.S. government's priorities in a given country. The integrated country strategy sets mission goals and objectives through a coordinated and collaborative planning effort among State, USAID, and other U.S. government agencies with programming in country. The primary audiences for the integrated country strategy are the mission, bureaus, and interagency partners. As of February 2018, missions will complete an integrated country strategy every 4 years. Current integrated country strategies were updated in February 2018 through 2022.

[42] We selected 10 strategies out of 69 for review by using the following two criteria: (1) they were located in either State's Africa or East Asia and the Pacific bureaus, which were the first two bureaus to implement climate risk screenings; and (2) they were identified by USAID and DOD as climate change priorities or experiencing displacement due to natural disasters. The 10 strategies selected were Cambodia, Ethiopia, Indonesia, Kenya, Mali, Philippines, Senegal, Timor-Leste, Uganda, and Vietnam.

migration. Of the regional bureau strategies we reviewed, we found that one, the Bureau for East Asian and Pacific Affairs, identified climate change as a driver of migration as a challenge or risk in its region. For example, the strategy states that climate change is becoming increasingly disruptive, potentially increasing migration due to rising sea levels. None of the other six regional bureau strategies we reviewed identified the nexus of climate change and migration as a risk or challenge. However, five regional bureaus identified climate change as a risk or challenge and one identified migration as a risk or challenge.

- USAID. USAID also requires the integration of climate risk management into all country or regional development cooperation strategies drafted since October 1, 2015.[43] Missions must document in a climate change appendix to the strategy any climate risks they identified and how they considered climate change in their strategy. As of August 2018, USAID had completed five country or regional development cooperation strategy updates initiated since October 1, 2015—Uganda, Tunisia, East Africa, Sri Lanka, and Zimbabwe—and all five included the required appendix.[44] Of the five updated strategies, three—Uganda, Tunisia, and East Africa—discuss the indirect effect of climate change on migration, among other issues. For example, Uganda's 2016-2021 country strategy states that increased frequency and duration of droughts is likely to be the most significant climate-related change in Uganda. The strategy also notes that droughts have affected, and

---

[43] According to USAID, climate risk management is the process of assessing, addressing, and adaptively managing climate risks. Climate risk management is required as part of the development of all new country or regional USAID strategies. Country Development Cooperation Strategies or Regional Development Cooperation Strategies define a mission's goal and objectives for an agreed-upon period of time, based on a given level of resources, and support State-USAID Joint Regional Strategies, Integrated Country Strategies, and the State-USAID Joint Strategic Plan. USAID's strategies are developed on a rolling basis, and generally span 5 year.

[44] There are 63 country and regional strategies. According to USAID officials, all new strategies initiated as of October 1, 2015, include a climate risk screening annex, as required. USAID officials stated that, as of July 2018, eight other missions are working on developing strategies that comply with the climate risk requirements: Afghanistan, El Salvador, Ethiopia, India, Liberia, Pakistan, Philippines, and the Ukraine.

will continue to affect, water resources, hydroelectricity production, and agriculture, among other sectors. As agriculture, forestry, and fisheries decline in Uganda, the strategy asserts that people will migrate to urban areas, leading to the formation of slums. We also reviewed USAID's nine regional development cooperation strategies, one of which—East Africa—had been updated since the requirement to include climate risk management. Of the other eight strategies that have yet to be updated, seven identified climate change as a challenge or risk and three identified climate change as a driver of migration as a challenge or risk. For example, the Southern Africa regional development cooperation strategy states that water scarcity, natural disasters, and other climate change related events will most likely increase migration throughout the region. Additionally, the Asia regional development cooperation strategy discusses the risks of climate change in urban areas. In Asia, the number of migrants seeking economic opportunities in urban centers is likely to increase. According to the strategy, migrants are moving into hazard-prone areas located along coastlines, flood plains, and other low-lying areas in many Asian primary and secondary cities—areas that experts predict will experience more frequent and intense storm surges, floods, and coastal erosion as a result of climate change.

State Lacks Clear Guidance on its Process for Assessing Climate Change-Related Risks

The requirement in Executive Order 13677 to assess climate change-related risks to agency strategies remains unchanged; however, State now lacks clear guidance on its process for assessing climate change- related risks to its integrated country strategies. [45] Specifically, State's 2016 guidance for developing integrated country strategies stated that all missions should assess the risk of climate change on their strategies goals

[45] Exec. Order No. 13,677, 79 Fed. Reg. 58,231 (Sept. 23, 2014).

and objectives[46] and included reference to the climate risk screening tool—a method that missions could use to assess climate change risks.[47] State issued new guidance to its missions in 2018, but this guidance does not include information on the process for assessing climate change-related risks to agency strategies. According to State officials, the 2018 guidance for integrated country strategies does not reference climate change risk assessments because, in September 2017, State decided that the strategies should not single out climate change risks in a separate appendix. State officials said this decision resulted, in part, from the new administration's shift in priorities on climate change. Officials also said that this decision reflects a new approach to risk management by State and that the missions could choose to include climate change and other potential risks in the general risk discussion section of their strategies. Officials from State's Office of U.S. Foreign Assistance Resources said that it is now up to each mission to decide whether a strategic objective may have a climate challenge. However, those missions that choose to include an assessment of climate change risks are not provided guidance on the process for doing so and there is no reference to the climate risk screening tool—or to climate change at all—in the 2018 guidance.

Executive Order 13677 directed State to incorporate climate-resilience considerations into decision making by assessing climate-related risks to agency strategies, among other things. Subsequently, a State cable from September 2016 further explained that State would implement the executive order's requirement by screening for climate risks as part of the process for drafting all new integrated country strategies. Additionally, the *Standards for Internal Control in the Federal Government* state that documentation is a necessary part of an effective internal control system.[48] If management determines that a principle is not relevant, management

[46] In 2016, State required missions to perform a climate change risk assessment, and if they determined that one or more goals and objectives had a medium or high risk from climate change, the missions had to document this climate risk screening process in an annex.

[47] State's screening tool outlines four steps for climate-risk screening. These steps are: (1) identify priority sectors and regions; (2) assess potential climate change impacts and risks; (3) adjust the strategy, if needed; and (4) document the climate risk screening process, if needed.

[48] GAO, *Standards for Internal Control in the Federal* Government, GAO-14-704G (Washington, D.C.: September 2014).

must support that determination with documentation that includes the rationale of how, in the absence of that principle, the associated component could be designed, implemented, and operated effectively.

Because State lacks clear guidance on its process for assessing climate change-related risks to its integrated country strategies, it is less likely that the current round of strategies will include the assessment of climate-related risks. It is also possible that those missions that choose to conduct climate change risk assessments will not do so in a consistent manner. Such assessments might identify climate change as a driver of migration, as at least one previous assessment did under the 2016 guidance. Thus, without clear guidance, missions may not examine climate change as a risk to their strategic objectives and could miss opportunities to improve the climate resilience of foreign assistance activities.

State, USAID, and DOD Have Been Involved in Various Climate Change Related Activities, but None Were Focused Specifically on Migration, and Their Participation Has Declined

For fiscal years 2014 through 2017, State, USAID, and DOD had some activities that could potentially address climate change as a driver of migration, although none of these activities specifically focused on the issue. For example, USAID has climate change adaptation activities, but to date migration has not been a focus of this programming. With the shift in priorities related to climate change in fiscal year 2017, agencies have reduced some of these activities.

State Activities

State's offices that are focused on the issues of climate change (State/OES) and migration (State/PRM) have participated in multilateral activities related to climate change as a driver of migration and funded adaptation and other activities related to the issue. State officials said that the agency does not, however, have any activities that specifically address migration due to climate change or environmental factors.

Multilateral Efforts

Table 1. Department of State (State) Participation in Multilateral Activities Related to Climate Change and Migration

Multilateral Activity	Description of Multilateral Activity and State Participation
Friends of the Nansen Initiative	The Friends of the Nansen Initiative was a group of countries and organizations associated with the Nansen Initiative. The Nansen Initiative aimed to develop an agenda for protecting people displaced across borders by disasters and the effects of climate change. The United States joined the group of friends in 2014 to engage on issues related to population movements linked to climate change. The United States was not a member of the initiative itself.
Platform on Disaster Displacement	The Platform on Disaster Displacement succeeded the Friends of the Nansen Initiative in May 2016. The platform addresses the protection needs of people displaced across borders in the context of disasters and climate change. As of September 2018, the United States is not participating in the Platform on Disaster Displacement.
Paris Agreement	The Paris Agreement aims to strengthen the global response to the threat of climate change. In a decision accompanying the Paris Agreement, the parties to the UN Framework Convention on Climate Change[a] requested that the Executive Committee of the Warsaw International Mechanism on Loss and Damage establish a task force on human displacement related to the adverse impacts of climate change. In August 2017, the United States formally announced its intent to withdraw from the Paris Agreement.[b]
Warsaw International Mechanism for Loss and Damage	The Warsaw International Mechanism for Loss and Damage was created to address loss and damage associated with impacts of climate change in vulnerable developing countries. The Warsaw International Mechanism for Loss and Damage's Executive Committee guides the implementation of the

Table 1. (Continued)

Multilateral Activity	Description of Multilateral Activity and State Participation
	mechanism. The Executive Committee's Task Force for Displacement is developing recommendations for approaches to address displacement related to the adverse impacts of climate change. The United States sits on the Warsaw International Mechanism for Loss and Damage's Executive Committee but does not participate in the Task Force.
The UN Framework Convention on Climate Change—Cancun Adaptation Framework	The Cancun Adaptation Framework provides broad guidance to strengthen climate-related adaptation, including disaster risk reduction strategies, among other things. The United States negotiates on issues created by the Cancun Adaptation Framework, including but not limited to: National Adaptation Plans, Loss and Damage, and the Adaptation Committee.
The UN Framework Convention on Climate Change—Nairobi Work Program	The Nairobi Work Program seeks to improve the understanding and assessment of the impacts of climate change and strengthen the capacity of countries to make informed decisions on practical adaptation actions. The United States participates in Nairobi Work Program activities.
The UN Global Compact for Migration	The UN Global Compact for Migration is expected to be the first intergovernmental agreement to cover all dimensions of international migration, including the humanitarian, developmental, and human rights aspects. Among other issues, the draft compact promotes initiatives related to climate change and migration, requests investment in programs that minimize climate change as a driver of migration, and calls for the strengthening of information on migration including the adverse effects of climate change. In December 2017, the United States announced its withdrawal from negotiations for the UN Global Compact for Migration.

Source: GAO analysis of Department of State and multilateral activity information. | GAO-19-166.

[a] The UN Framework Convention on Climate Change was adopted in 1992. Currently, the United States is one of 197 parties to the convention. The convention's objective is to stabilize greenhouse gas concentrations in the atmosphere at a level that would prevent dangerous anthropogenic (man- made) interference with the climate system.

[b] In June 2017, the President announced his intention to withdraw the United States from the Paris Agreement. Under the agreement, the earliest effective date of withdrawal for the United States is in 2020.

State has participated in multilateral activities related to climate change and migration. With the shift in priorities related to climate change in fiscal year 2017, the United States has disengaged from some of these multilateral activities (see table 1).

Funding Efforts

In addition to State's participation in the multilateral activities described in table 2, State has provided funding for activities related to climate change and capacity building that address natural disasters. These activities may involve efforts potentially related to migration. For example, according to State:

- State provided about $2 million per year, between fiscal years 2014 and 2016, to the Intergovernmental Panel on Climate Change,[49] which analyzed the impacts of climate change on migration in its most recent assessment report. [50]
- State/PRM provided about $4 million, between fiscal years 2014 through 2018, for IOM's Migrants in Countries in Crisis Initiative, which provides guidelines to protect migrants in countries experiencing conflict or natural disasters. IOM provides training to countries on these guidelines. State/PRM officials said that this initiative is not specifically related to climate change and does not focus on specific types of disasters but does mention sudden-onset disasters. Officials also said that IOM tries to promote a climate change perspective in its trainings.

---

[49] According to State's Bureau of International Organization Affairs, as of fiscal year 2017, State is no longer providing funding for the Intergovernmental Panel on Climate Change.

[50] Adger, W.N., J.M. Pulhin, J. Barnett, G.D. Dabelko, G.K. Hovelsrud, M. Levy, Ú. Oswald Spring, and C.H. Vogel, *Human Security. In: Climate Change 2014: Impacts, Adaptation, and Vulnerability. Part A: Global and Sectoral Aspects. Contribution of Working Group II to the Fifth Assessment Report of the Intergovernmental Panel on Climate Change* [Field, C.B., V.R. Barros, D.J. Dokken, K.J. Mach, M.D. Mastrandrea, T.E. Bilir, M. Chatterjee, K.L. Ebi, Y.O. Estrada, R.C. Genova, B. Girma, E.S. Kissel, A.N. Levy, S. MacCracken, P.R. Mastrandrea, and L.L. White (eds.)]. (Cambridge University Press, Cambridge, United Kingdom and New York, NY, USA: 2014).

- State/OES provided about $78 million in adaptation funding from the Global Climate Change Initiative to eight projects during fiscal years 2014 through 2017.[51] (See appendix III for a description of all eight projects.) State/OES officials said that these projects help countries prepare for the impacts of climate change, potentially reducing the pressure to migrate. However, to these officials' knowledge, none of these projects directly supported activities related to migration. For example, State/OES provided a $4 million grant to the National Adaptation Plans Global Network. This network focuses on increasing the capacity of governments to identify and assess climate risks, integrate these risks in planning, develop a pipeline of projects to address these risks, identify and secure funding for projects, and track progress toward resilience targets. Adaptation activities occurred in over 35 countries.

With the shift in priorities related to climate change in fiscal year 2017, State discontinued some of these efforts. For example, funding for the Global Climate Change Initiative was not included in the President's budget request for fiscal year 2018. State/OES officials said that the agency does not plan to fund additional adaptation activities and has not requested additional funding for the activities. According to a State official, PRM had been in discussion with IOM to develop a project proposal that would have assisted the governments of Small Island Developing States in adapting their migration policies to account for challenges and opportunities associated with environmental degradation, ecosystem loss, climate change impacts, and natural disasters. State/PRM stopped further development of the proposal following the change in administrations.

[51] The Global Climate Change Initiative was established in 2010 to promote resilient, low-emission development, and integrate climate change considerations into U.S. foreign assistance. The initiative was divided into three main programmatic initiatives: (1) Adaptation assistance, (2) Clean Energy assistance, and (3) Sustainable Landscapes assistance.

Additionally, according to a State official, the department made some efforts at the end of the previous administration to develop a formal position on the topic of climate change as a driver of migration. For example, State drafted an internal document to help clarify its role in responding to the humanitarian aspects of sudden-onset and slow-onset climate events. This initial work stopped under the current administration.

USAID Activities

USAID officials said that, with respect to the agency's climate-related programming, its climate change adaptation programming was the most likely to include activities related to migration or displacement, although a broad swath of USAID development programming has the potential to build host country resilience. Officials stated that, to date, migration has not been a primary motivation for the agency's climate-related or disaster assistance programming. However, officials said that, in a humanitarian crisis or under some economic conditions, development programming can reduce displacement or the pressure to migrate—such as by fostering greater resilience to drought or other adverse conditions—and that this is also true of climate-related programming. USAID also provides humanitarian assistance in response to natural disasters that displace people. Officials said that USAID recognizes the links between displacement and natural disasters, but that the agency does not have specific programs linking disaster assistance, migration, and climate change.

Adaptation Efforts

USAID identified about 250 activities that received adaptation funding from the Global Climate Change Initiative during fiscal years 2014 through 2016.[52] Our analysis of the descriptions of these activities determined that

[52] Beginning in fiscal year 2017, State and USAID no longer allocated dedicated funding to climate change adaptation, as they had under the Global Climate Change Initiative.

none directly mentioned any efforts specifically related to migration.[53] Officials emphasized that the connection between climate change and migration tends to be indirect and shaped by other more immediate factors. USAID's data on activities that received adaptation funding identified 38 beneficiary countries, as well as activities described generally as implemented at the regional or global level.[54] For activities where USAID's data identified a specific region, most activities were located in Africa followed by Asia and Latin America and the Caribbean.[55]

Examples of the types of activities that received adaptation funding from the Global Climate Change Initiative during fiscal years 2014 through 2016 include:

- The Mali Climate Change Adaptation Activity, which aims to build resilience to current climate variability and increase resilience to longer-term climate change effects. This activity is also working to strengthen the capacity of Mali's meteorological agency to provide improved climate information as well as to incorporate climate considerations into local-level planning. The total estimated cost is about $13 million over 5 years.[56]

---

[53] We conducted an automated review of the activity description fields provided by USAID for terms related to migration, see appendix I for more information. Agency officials said that USAID records funding and results data according to the original purposes of the funding up until the date of obligation. According to officials, USAID does not have a way of retrieving from its data systems whether a project includes activities related to migration or displacement (or any other topic), unless the topics were part of the original intent or purpose of the funding. Furthermore, officials indicated that there are no project or activity documents that would aggregate funding levels and funding source, scope of work, and assessment of impact specifically on migration or displacement (or any other topic), unless the topics were part of the original intent or purpose of the funding.

[54] The beneficiary country field was empty—either not specifying a specific country or indicating that the activity was a regional or global activity—for 55 of the activities in USAID's data.

[55] USAID's data had specific region information for 160 of the 254 activities in the data. Of those activities where a specific region was identified, 76 activities were located in Africa (about 48 percent), 48 were located in Asia (about 30 percent), and 36 were located in Latin America and the Caribbean (about 23 percent)

[56] According to USAID, total estimated cost may not be fully funded depending on availability of funds, the final amount negotiated with the implementer, changes in country needs or the Country or Regional Development Cooperation Strategy, and or, other issues. Further, USAID said that activities often address multiple development objectives; therefore, not all of the total estimated cost can be attributed to climate adaptation funding.

- The activity for Climate-Resilient Ecosystems and Livelihoods, which ended in September 2018, aimed to increase Bangladesh's resilience to natural hazards by working with community-based organizations, government ministries, and technical agencies. This activity provided technical assistance to the Government of Bangladesh and local communities to improve ecosystem conservation and resilience capacity. The total estimated cost was about $33 million in funding over 6 years.
- The activity for Pastoralist Areas Resilience Improvement through Market Expansion, which aims to support pastoralists in Ethiopia via expansion of markets and long-term behavior change (see fig. 3). USAID officials cited this activity as an example of adaptation efforts that indirectly address the issue of climate change as a driver of migration. The activity has three interrelated objectives: increasing household incomes, enhancing resilience, and bolstering adaptive capacity to climate change among pastoral people in Ethiopia. An evaluation of the activity found that migration is a coping strategy for dealing with climate shocks, although participants said that drought is becoming more frequent, placing a severe strain on traditional coping mechanisms, such as migration and selling cattle, and that permanent migration is not a preferred strategy. The total estimated cost is about $60 million in funding over 6 years.

With the shift in priorities related to climate change, funding for USAID's climate change adaptation activities has decreased. Missions may continue to fund their adaptation activities with discretionary funds or other earmarked, sector funding, provided the activities further the funding source's objective, according to USAID. For example, in some cases, missions are using Water sector funding to continue some of their adaptation work. USAID also said that among the agency's goals are to increase the resilience of USAID partner countries to recurrent crises, including climate variability and change.

Source: Mercy Corps. | GAO-19-166.

Figure 3. Ethiopia Pastoralist Areas Resilience Improvement through Market Expansion Activity is Relevant to the Issue of Climate Change as a Driver of Migration.

Humanitarian Aid and Disaster Assistance Efforts

In addition to USAID's climate change adaptation programming, USAID/OFDA and USAID/FFP provide emergency humanitarian assistance to people affected by sudden-onset disasters—such as hurricanes and floods—and slow-onset and extended disasters, including droughts and conflicts. Some of this assistance helps people who have been displaced by disaster. USAID officials stated that although disasters cause mainly temporary displacement, the relationship among humanitarian assistance, climate change, and migration is very complex and depends on both climatic and non-climatic factors. USAID/OFDA responded to 267 disasters from fiscal year 2014 through June 2018, according to agency data. For example, USAID/OFDA responded to the effects of Hurricane Matthew in Haiti in October 2016, as seen in figure 4, including helping temporarily displaced people.

Source: Mike Lambright, United States Agency for International Development (USAID)/Office of U.S. Foreign Disaster Assistance (OFDA). | GAO-19-166.

Figure 4 USAID Aerial View of Hurricane Matthew Damage in Haiti, October 2016, Which Temporarily Displaced People.

DOD Activities

DOD assists in the U.S. government response to overseas disasters, including helping people displaced by such disasters, regardless of the cause of the disaster. These efforts are not specific to climate change as a driver of migration. For example, officials from DOD's geographic combatant commands said that, to the extent they address climate change, migration is not a focus of those efforts and they view migration as caused by security and economic issues.

Between fiscal years 2014 and 2018, Congress has appropriated to DOD between $103 and $130 million per year for Overseas Humanitarian, Disaster, and Civic Aid. Officials said that the geographic combatant commands use most of this funding for steady state humanitarian assistance related to health, education, basic infrastructure, and disaster preparedness with a smaller amount set aside for immediate disaster assistance although that varies based on emergency requirements. DOD officials said that they have not seen any changes to this funding or associated activities with the change of administrations in fiscal year 2017. DOD officials we spoke with also emphasized that USAID/OFDA is the lead agency for the U.S. government's response to disasters overseas. USAID/OFDA formally requested DOD support on about 10 percent of the foreign disaster assistance provided by USAID/OFDA, according to USAID data for fiscal year 2014 through June 2018 and DOD officials. DOD assistance is typically provided for the largest, most complex disasters, according to agency officials.

According to a July 2015 assessment conducted by the geographic combatant commands, while their activities vary, each command works with partner nations to increase their abilities to reduce the risks and effects from environmental impacts and climate-related events, including severe weather and other hazards.[57] For example, in the report, U.S. Southern Command stated that it had requested funding to pre-position assets for when a severe storm threatens Haiti to be able to respond immediately to a

[57] Department of Defense, National Security Implications of Climate-Related Risks and a Changing Climate (Washington, D.C.: July 2015.

potential disaster. U.S. Southern Command officials said that they work with partner nations to encourage residents experiencing extreme weather to remain where they are because it is easier to provide help to people who stay in one place. Officials from U.S. Southern Command and U.S. Africa Command also said that the major factors driving migration in their regions are security and economic issues.

Interagency Forums

State, USAID, and DOD have participated in interagency forums regarding climate change, which may have addressed its effects on migration. With changes to priorities regarding climate change in fiscal year 2017, these forums have been disbanded or are not meeting.

- The Council on Climate Preparedness and Resilience. The Council on Climate Preparedness and Resilience, of which State, USAID, and DOD were members, was established to facilitate the integration of climate science in policies and planning of government agencies, including by promoting the development of climate change related information, data, and tools, among other things. Additionally, the council was to develop, recommend, and coordinate interagency efforts on priority federal government actions related to climate preparedness and resilience. According to State officials, the council began working with the National Security Council and other agencies to facilitate greater interagency cooperation on adaptation. In addition, a task force on the council was discussing the federal role in addressing displacement related to climate change. The council was disbanded when Executive Order 13783 revoked Executive Order 13653, which had established the council.[58]
- The Working Group on Climate-Resilient International Development. The Working Group on Climate-Resilient

[58] Exec. Order No. 13,783, 82 Fed. Reg. 16,093 (Mar. 28, 2017

International Development, of which State and USAID were members, was established by Executive Order 13677 and placed under the Council on Climate Preparedness and Resilience.[59] The working group's mission includes developing guidelines for integrating considerations of climate-change risks and climate resilience into agency strategies, plans, programs, projects, investments, and related funding decisions, among other things. Additionally, the working group was tasked with facilitating the exchange of knowledge and lessons learned in assessing climate risks to agency strategies, among other things. USAID officials said that the working group had not discussed climate change as a driver of migration. While the working group has not been formally disbanded, it has not met since at least November 2017 according to USAID.

- The Climate and National Security Working Group. The Climate and National Security Working Group, of which State, USAID, and DOD were members, was established by the 2016 presidential memorandum.[60] The chairs of the working group were to coordinate the development of a strategic approach to identify, assess, and share information on current and projected climate-related impacts on national security interests and to inform the development of national security doctrine, policies, and plans, among other things. According to the memorandum, the working group was to provide a venue for enhancing the understanding of the links between climate change- related impacts and national security interests and for discussing opportunities for climate mitigation and adaptation activities to address national security issues. This working group was disbanded when Executive Order 13783 revoked the 2016 presidential memorandum, which had established the working group.

---

[59] Exec. Order No. 13,677, 79 Fed. Reg. 58,231 (Sept. 23, 2014).

[60] Presidential Memorandum, Climate Change and National Security (Sept. 21, 2016).

Conclusion

State, USAID, and DOD assessments and activities have not focused specifically on the nexus of climate change and migration. State did identify migration as a risk of climate change in at least one of its climate change risk assessments for the department's country strategies. However, State now lacks clear guidance on its process for assessing climate change-related risks to its integrated country strategies. State's current guidance for these country strategies no longer mentions a climate change risk assessment and does not provide missions with information about the climate risk screening tool that can be used to conduct such an assessment. As such, missions are less likely to examine climate change as a risk to their strategic objectives, or to do so in a consistent manner, and thus may not have the information they would need to identify migration as a risk of climate change. By clearly documenting and providing guidance on how to assess the risk of climate change, State would ensure that the department examines the potential risks of climate change on its foreign assistance activities.

Recommendation for Executive Action

We are making the following recommendation to State:

The Secretary of State should ensure that the Director of the Office of U.S. Foreign Assistance Resources provides missions with guidance that clearly documents the department's process for climate change risk assessments for integrated country strategies. (Recommendation 1)

Agency Comments

We provided a draft of this product to State, USAID, and DOD for review and comment. State provided written comments, which we have reprinted in appendix IV. In its comments, State did not oppose the recommendation and noted that the agency will update its integrated country strategy guidance by June 30, 2019 to inform missions that they have the option to include an annex on climate resilience, as well as other topics. However, State also indicated that the agency will begin working with stakeholders to consider whether to recommend that the Secretary of State ask the President to rescind Executive Order 13677: Climate-Resilient International Development.

USAID also provided written comments, which we have reprinted in appendix V. In its letter, USAID provided some additional information about its programs and its proposed transformation effort. USAID and DOD provided technical comments, which we incorporated as appropriate.

We are sending copies of this chapter to the appropriate congressional requesters, Secretary of State, the Administrator of USAID, and the Secretary of Defense.

David Gootnick
Director, International Affairs and Trade

Brian J. Lepore
Director, Defense Capabilities and Management

List of Requesters

The Honorable Sheldon Whitehouse
United States Senate

The Honorable Dianne Feinstein
United States Senate

The Honorable Elizabeth Warren
United States Senate

The Honorable Edward J. Markey
United States Senate

The Honorable Jeffrey A. Merkley
United States Senate

APPENDIX I: OBJECTIVES, SCOPE, AND METHODOLOGY

This chapter (1) describes executive branch actions related to climate change and migration from fiscal years 2014 through 2018; (2) examines the extent to which the Department of State (State), the U.S. Agency for International Development (USAID), and the Department of Defense (DOD) have discussed the potential effects of climate change on migration in their plans and risk assessments; and (3) describes State, USAID, and DOD activities, if any, that are related to climate change and global migration. We chose fiscal years 2014 through 2018 as our time frame based on our review of recent executive orders related to climate change. We selected State, USAID, and DOD because the agencies' missions of diplomacy, development, and defense provide the foundation for promoting and protecting U.S. interests abroad.

To describe executive branch actions related to climate change and migration from fiscal years 2014 through 2018, we reviewed documents that reflect priorities of the previous and current administrations. Specifically, we reviewed budget requests and enacted appropriations between fiscal years 2014 through 2018 for funding priorities related to climate change and U.S. foreign assistance. In addition, we reviewed executive actions and executive branch strategies that applied to State, USAID, and DOD between fiscal years 2014 through 2018 for executive and national security priorities related to climate change. For example, we reviewed the current and previous national security strategies.

To examine the extent to which State, USAID, and DOD have discussed the potential effects of climate change on migration in their plans and risk assessments, we reviewed documents completed since fiscal year

2014 such as Agency Adaptation Plans, previously required by Executive Order 13653; country and regional strategies; and policy guidance. We also met with officials from these agencies to discuss these assessments. We reviewed State and USAID country and regional strategies that were required by Executive Order 13677 to be assessed for climate risk.[61] For State, we chose integrated country strategies from 10 missions by selecting the countries that fit the following two criteria: (1) they were located in either State's Africa or East Asia and the Pacific bureaus, which were the first two bureaus to implement climate risk assessments; and (2) they were identified by USAID and DOD as climate change priorities or experiencing displacement due to natural disasters. To obtain a broader perspective, we also reviewed three functional bureau strategies and seven regional bureau strategies. For USAID, we examined the five country and regional strategies that were required to include a climate risk assessment at the time of our review: Uganda, Tunisia, East Africa, Sri Lanka, and Zimbabwe. We also reviewed all nine USAID regional strategies. For both State and USAID, we reviewed the selected strategies by searching for information related to migration and climate change. To determine whether State clearly documents the department's current climate risk assessment process for integrated country strategies, we compared State's 2018 guidance for developing integrated country strategies with standards related to documentation in *Standards for Internal Control in the Federal Government* and previous State guidance issued in 2016, which was created in response to Executive Order 13677's requirements to assess climate change risks to strategies, among other things.[62]

To describe State, USAID, and DOD activities, if any, that are related to climate change and global migration, we requested that agencies identify and provide information on efforts that were active between fiscal years 2014 and 2017.

---

[61] Exec. Order No. 13,677, 79 Fed. Reg. 58,231 (Sept. 23, 2014).

[62] GAO, Standards for Internal Control in the Federal Government, GAO-14-704G (Washington, D.C.: September 2014).

- State officials said that the agency received adaptation funding from the Global Climate Change Initiative during this period so we requested a list of projects from State's Bureau of Oceans and International Environmental and Scientific Affairs (State/OES). State/OES provided all adaptation projects funded with State/OES funds in fiscal year 2014 or later. We reviewed the summaries of those eight projects for discussion of migration. Because no State adaptation projects specifically mentioned migration, for the purposes of this chapter we chose illustrative examples to provide context for the types of projects the agency has funded. State/OES and the Bureau for Population, Refugees, and Migration also discussed international initiatives in which they have participated that have activities relevant to climate change as a driver of migration.
- USAID officials from the Bureau for Economic Growth, Education, and Environment identified activities that received adaptation funding from the Global Climate Change Initiative as the most likely to be relevant to these issues.[63] The agency then provided us with data for about 250 activities from its annual operational plans for fiscal years 2014 through 2016, the 3 years during the period we reviewed in which it received adaptation funding. USAID identified these activities based on whether the agency had tagged them in its plans as having an "adaptation key issue." USAID excluded projects that had planned attributions to the adaptation key issue of less than $250,000 in a given fiscal year, as well as certain other activities such as those that focused on project support. We then conducted an automated review of the activity description fields provided by USAID for terms related to migration and other descriptive information such as locations of activities. Because no USAID adaptation activities specifically

[63] The Global Climate Change Initiative was established in 2010 to promote resilient, low-emission development, and integrate climate change considerations into U.S. foreign assistance. The initiative was divided into three main programmatic initiatives: (1) Adaptation assistance, (2) Clean Energy assistance, and (3) Sustainable Landscapes assistance.

mentioned migration, for the purposes of this chapter we chose illustrative examples to provide context for the types of activities the agency has funded.

- DOD officials we met with did not identify any specific activities related to climate change as a driver of migration. DOD officials from the Assistant Secretary of Defense for Special Operations and Low Intensity Conflict and the geographic combatant commands[64] generally discussed DOD activities related to humanitarian assistance and disaster response as most relevant to our inquiry. Because DOD works in coordination with USAID's Office of U.S. Foreign Disaster Assistance on disaster assistance we also reviewed USAID data on its disaster response activities during this period.

We determined that the USAID and State adaptation project data and USAID disaster assistance data were sufficiently reliable for the purposes of describing these efforts.

We also obtained information on the effects of any changes in priorities related to climate change on these agency activities, as well as on interagency coordination. We reviewed relevant presidential budget requests and executive actions. In addition, we met with officials from State, USAID, and DOD to obtain information on whether changes in government priorities related to climate change affected their activities.

We conducted this performance audit from October 2017 to January 2019 in accordance with generally accepted government auditing standards. Those standards require that we plan and perform the audit to obtain sufficient, appropriate evidence to provide a reasonable basis for our findings and conclusions based on our audit objectives. We believe that the evidence obtained provides a reasonable basis for our findings and conclusions based on our audit objectives.

---

[64] DOD's geographic commands include: U.S. Africa Command, U.S. Central Command, U.S. European Command, U.S. Northern Command, U.S. Indo-Pacific Command, and U.S. Southern Command.

APPENDIX II: REGIONAL FOCUS ON CLIMATE CHANGE AS A DRIVER OF GLOBAL MIGRATION

This appendix provides a review by region of observed and projected climate change effects, migration trends, and challenges in stability and security. Multiple sources we used for this overview make a connection between climate change and such events as rising sea levels, higher temperatures, and an increase in the number and severity of extreme weather events.[65] The following regions are discussed: Asia, South America, the Arctic, Sub-Saharan Africa, the Middle East and North Africa, Oceania, and Central America and the Caribbean. We have provided an overview for each region and a focus on one country or territory in the region.

To provide this overview, we reviewed information from a number of sources.[66] First, we reviewed reports from international organizations focused specifically on climate change or migration: the Intergovernmental Panel on Climate Change (IPCC);[67] the United Nations Environment Programme (UNEP); the International Organization for Migration (IOM); the United Nations High Commissioner for Refugees (UNHCR); and the

---

[65] See previous GAO work related to the uncertainties in projecting the effects of climate change, GAO, Climate Information: A National System Could Help Federal, State, Local, and Private Sector Decision Makers Use Climate Information, GAO-16-37 (Washington, D.C.: November 23, 2015). According to a 2012 National Research Council report on climate models, the fundamental science of greenhouse gas-induced climate change is simple and compelling. However, genuine and important uncertainties remain, such as how clouds affect the climate system, and these uncertainties need to be considered in developing scientifically based strategies for societal response to climate change— especially those related to "downscaled" climate information. For more information, see National Research Council, A National Strategy for Advancing Climate Modeling (Washington, D.C.: 2012).

[66] For domestic climate science information, GAO relies on the National Academies of Science, Engineering, and Medicine and the U.S. Global Change Research Program. According to these sources, limiting the federal government's fiscal exposure to climate change risks will be challenging no matter the outcome of efforts to reduce emissions, in part because greenhouse gases already in the atmosphere will continue altering the climate system for many decades.

[67] The IPCC is the UN body for assessing the science related to climate change. We reviewed information from IPCC's Fifth Assessment Report (AR5), which is the organization's most comprehensive assessment of scientific knowledge on climate change since 2007. The report was released in four parts between September 2013 and November 2014. The IPCC's Sixth Assessment Report will also be released in four parts between April 2021 and April 2022.

African, Caribbean, and Pacific States (ACP) Observatory on Migration.[68] Second, we reviewed relevant reports from international and regional organizations, including a variety of organizations within the United Nations, the World Bank, regional development banks, the European Union, and others. Third, we reviewed relevant public documents from U.S. government agencies, including the Department of Defense, the U.S. Agency for International Development (USAID), and the United States Institute of Peace (USIP). Fourth, we reviewed academic sources, research institutions, and documents from the relevant country's national government.

The information about each country or territory also includes the following indices and statistics, chosen to illustrate underlying conditions that impact climate resilience, migration, and stability:

- Total Population: The total population of a country, area, or region.[69]
- Fragile States Index: An annual ranking of 178 countries based on factors that impact stability, with 1 being the least stable, and 178 the most stable. The index accounts for economic, political, and social factors, among others.[70] According to a variety of sources, fragile or unstable states may have limited resilience to cope with the effects of climate change, including a potential increase in migration.
- Human Development Index: A composite index covering three dimensions: life expectancy at birth, mean and expected years of schooling, and gross national income per capita. The Human Development Index classifies countries as Low, Medium, High, or

---

[68] The ACP Observatory on Migration is an initiative to study South-South migration and development. Financial support for the initiative comes from the European Union, Switzerland, the United Nations Population Fund, and IOM.

[69] Country level statistics for total population from United Nations Development Program, *Human Development Indices and Indicators: 2018 Statistical Update* (New York, NY, 2018). Total population for Greenland from Population, total Data (2017) Retrieved from https://data. worldbank.org/indicator/SP.POP.TOTL?locations=GL.

[70] Fund for Peace, *2018 Fragile States Index* (Washington, D.C., 2018).

Very High Human Development.[71] Household characteristics and economic conditions may be factors for individuals in deciding whether to migrate.

- Gross Domestic Product (GDP) per Capita: The sum of gross value added by all resident producers in the economy divided by the total population.[72] Economic conditions may be a factor for people deciding whether to migrate or stay in their country of origin.
- Remittances as Percent of GDP: The money international migrants transfer to recipients in their country of origin, expressed as a percentage of the origin country's GDP.[73] Sources agree that remittances support resilience in origin countries.
- Agriculture, Fishing, Forestry as Percent of GDP: A measure of the value added to an economy from the agricultural sector, which includes forestry, hunting, fishing, and the cultivation of crops and livestock, expressed as a percentage of the country's GDP.[74] Countries that depend on the agricultural sector may be vulnerable to the effects of climate change, according to the World Bank.
- Percent of Population in Cities: The population living in areas classified as urban according to criteria each country uses.[75] Today, more than half of the global population lives in cities. Migration, in some cases due to climate change, is an important driver of urban growth, according to IOM. Cities are also expected to face increasing risks from rising sea levels, flooding, storms, and other climate change effects.
- Net Migration Rate: A measure of the number of people leaving a country compared to the number of people entering a country, expressed as a number per 1,000 people.[76]

---

[71] United Nations Development Program (2018)

[72] Country level statistics for GDP per Capita from United Nations Development Program (2018). GDP per Capita for Greenland, from GDP per Capita (current US$) Data. (2017). Retrieved from https://data. worldbank.org/indicator/NY.GDP.PCAP.CD.

[73] United Nations Development Program (2018).

[74] Agriculture, forestry, and fishing, value added (percent of GDP) Data. (2017). Retrieved from https://data.world bank.org/indicator/nv.agr.totl.zs.

[75] United Nations Development Program (2018).

[76] United Nations Development Program (2018).

ASIA

The effects of climate change in Asia may impact migration and stability according to the Intergovernmental Panel on Climate Change (IPCC) and the Asian Development Bank (ADB). In coastal areas, effects of climate change include rising sea levels, storm surges, and others. Receding glaciers in mountanous areas may also cause flooding, and monsoons in a warmer climate may be more severe. Heat extremes and more rainfall are a particular concern in Southeast Asia. Changes in precipitation and drought in Asia may exacerbate food security challenges, and contribute to people deciding to migrate. Increases in migration, partly stemming from the effects of climate change in surrounding rural areas, may put pressure on existing urban infrastructure. Rural migrants may settle in informal communities on the outskirts of cities, areas that have little resilience to natural disasters. Although the World Bank and others agree that climate change largely causes internal migration, some evidence shows that the impact of climate change contributes to cross-border migration in Asia.[77]

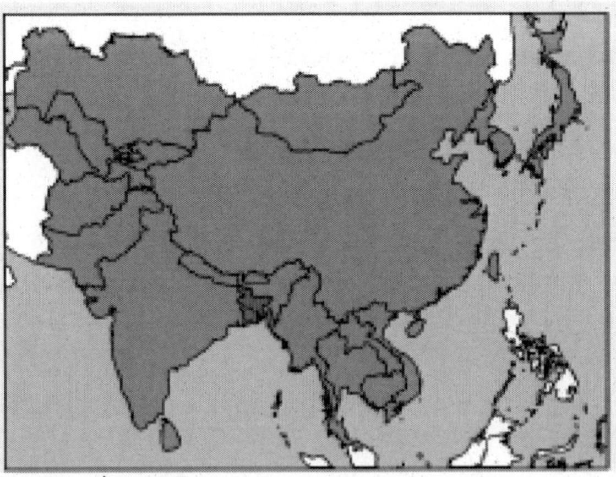

Source: Map Resources. | GAO-19-166.

[77] The Nansen Initiative, Global Consultation Conference Report (Geneva, October 12-13, 2015).

Large numbers of migrants, along with other destabilizing factors, may contribute to instability and conflict, according to the IPCC. The effects of climate change on livelihoods, for example, could increase migration, strain governance, and contribute to conflict as a result. Bangladesh is one example where decreased yields from agriculture and fisheries have contributed to migration to the country's coastal cities, which face their own climate change challenges.

BANGLADESH: CLIMATE CHANGE, MIGRATION, STABILITY AND SECURITY

Total Population
164,700,000
Fragile States Index
#32 out of 178
A lower number indicates greater instability.
Human Development Index
Medium
Gross Domestic Product (GDP)
per Capita
$3,524
Remittances as % of GDP
5.4
Agriculture, Fishing, Forestry as % of GDP
13.4
% Population in Cities
35.9
Net Migration Rate per 1000
people
-3.2

Observed and Projected Effects of Climate Change

Bangladesh's high population density and geography make the country susceptible to the effects of climate change, according to the World Bank, and others. Bangladesh's coasts and river banks are vulnerable to sudden-onset events such as tropical cyclones and flooding. Cyclone Aila in 2009, for example, caused widespread flooding in the southern coastal areas of Bangladesh and impacted millions of people. The storm washed away embankments that protected coastlines and caused severe damage to crops and livelihoods. Tropical Cyclone Mora in 2017 damaged thousands of homes and displaced an estimated 200,000 people. Increases in the number and intensity of tropical cyclones, which some predict will occur in a warmer climate, could have severe impacts on homes, livelihoods, and food security.[78]

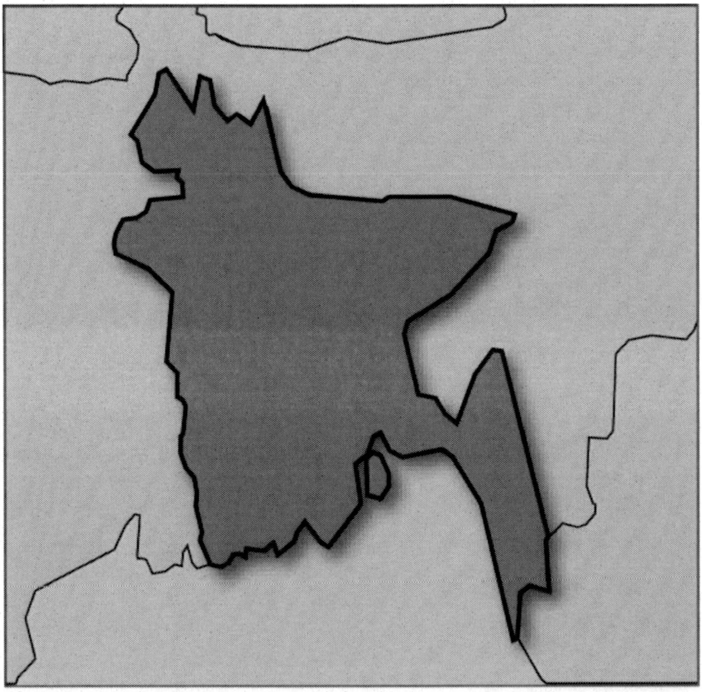

Source: Map Resources. | GAO-19-166.

Map of Bangladesh.

[78] See previous GAO work related to food assistance in Bangladesh, GAO, International Food Assistance: Agencies Should Ensure Timely Documentation of Required Market Analyses and Assess Local Markets for Program Effects, GAO-17-640 (Washington, D.C.: July 13, 2017).

Bangladesh also experiences many slow-onset climate change events, such as rising sea levels and increasingly severe droughts, which are projected to intensify with climate change. Bangladesh would lose an estimated 17.5 percent of its land if the sea level rose 1 meter, as the International Organization for Migration (IOM) has reported. Projected changes in precipitation levels could cause drought and food insecurity in the northwest and salt-water intrusion could reduce crop yields in the southwest.

Migration Trends

Migration is a common adaptation strategy to climate change in Bangladesh, according to the ADB. For example, some farmers have adapted to salt water intrusion and destroyed crops by switching to salt-tolerant rice production or shrimp cultivation. Others have migrated, often to Bangladesh's cities to find work less dependent on agriculture. Many new migrants to Bangladesh's cities live in informal settlements that lack the resilience to withstand sudden-onset climate events. The capital city, Dhaka, is a common destination for migrants displaced by salt-water intrusion, flooding, and river erosion, according to IOM. Dhaka, like many coastal cities in South Asia, is located on a low-lying riverbank and faces increasing risks of extreme flooding. For example, past floods in Dhaka have destroyed homes and contaminated drinking water, creating significant health hazards. In some cases, individuals migrate to cities temporarily for work and return home after the agricultural off season ends. Bangladeshis also provide a significant number of labor migrants to the Gulf States and Malaysia. Remittances from international migrants represent 5.4 percent of the country's GDP, and may help to support resilience to climate change, according to IOM, and others. These migration trends may intensify in the future. One study estimates 9.6 million people will migrate from 2011 to 2050 due to the effects of climate change.[79]

---

[79] Refugee and Migratory Movements Research Unit and the Sussex Centre for Migration Research, *Climate Change-related Migration in Bangladesh* (Climate and Development Knowledge Network, April 2013).

Challenges in Stability and Security

Migration due to climate change is cited as a potential destabilizing factor in Bangladesh by ADB, and others. The low-income population in Bangladesh is dependent on agriculture, making the effects of climate change—including impacts on food security—a particular concern. By 2030, these effects on livelihoods and food security could increase the poverty rate in Bangladesh by 15 percent, as the IPCC has reported.

Given the proximity of Bangladesh to India, some individuals may also choose to cross the border. Increased migration to India is a potential concern, according to some sources, as India may not have the resources to absorb large numbers of Bangladeshi migrants.[80]

Source: The UN Migation Agency (IOM) Abir Andullah.

A flooded house in Bangladesh.

[80] The CNA Corporation, *National Security and the Threat of Climate Change* (Alexandria, VA: 2007); and Population Council, "Effects of Future Climate Change on Cross-Border Migration in North Africa and India," Population and Development Review, Vol. 36, No. 2 (2010).

SOUTH AMERICA

The effects of climate change in South America vary by region, according to the the Intergovernmental Panel on Climate Change (IPCC) and International Organization for Migration (IOM), as well as potentially impacting migration and stability. On the coast, risks include sea level rise, depletion of fisheries, and coral reef bleaching, according to IOM.

Source: Map Resources. | GAO-19-166.

Coastal cities with growing populations are particularly vulnerable. Melting glaciers in the Andean mountain region, and increased rainfall are expected to change the distribution of water resources, and impact food production as global demand for food is growing. Desertification and land degradation, complicated by the effects of climate change, are contributing to migration from rural areas to cities in South America, as IOM has reported. An estimated 77 percent of people living in high risk areas in South America are located in cities, according to IOM. IOM predicts that as these people feel the effects of sea level rise and water scarcity, they will migrate from the large coastal cities to smaller urban areas. While South America has experienced economic growth in the last decade, poverty rates remain high, and the effects of climate change, including possible migration, may exacerbate inequalities, putting further pressure on cities to meet the needs of their populations. Water security in particular is expected to disproportionaly impact low-income communities, according to the IPCC. For example, in Brazil, drought in the northeast may increase migration to southern cities that are facing rising sea levels and landslides, with consequences for food, water, and energy security.

Brazil: Climate Change, Migration, Stability and Security

Total Population

209,300,000

Fragile States Index

#106 out of 178

A lower number indicates greater instability.

Human Development Index High

Gross Domestic Product (GDP)

per Capita

$14,103

Remittances as % of GDP

0.1

Agriculture, Fishing, Forestry as % of GDP

4.6

% Population in Cities

86.3

Net Migration Rate per 1000 people

0.0

Observed and Projected Effects of Climate Change

Brazil's cities and rural regions may encounter a range of climate change effects, according to the IPCC and IOM. Rural areas, particularly in the northeast, could experience significant impacts from climate change partly due to poverty rates, and historical vulnerability to drought. Higher temperatures are expected to affect crop yields and household incomes, especially for low-income communities. In northeastern Brazil, temperatures are expected to increase and rainfall to decrease. The northeast could see a 22 percent reduction in precipitation by 2100, according to IPCC projections. Brazil's coastal areas, including cities, are also vulnerable to rising sea levels, heavy precipitation, flooding, and landslides. The vast majority of Brazil's population, about 86 percent, lives in cities, many in coastal areas, according to the United Nations Development Program. As their populations have grown, urban areas have extended out. This urban growth in Brazil's megacities has caused further increases in temperature, rainfall, and landslides. For example, current levels of urbanization in the metropolitan area of Sao Paulo may already be responsible for the 2°C warming observed in the city over the last 50 years, as well as the rise in extreme rainfall, according to the IPCC. The metropolitan area is expected to extend its area 38 percent by 2030. Multiple studies of the effects of urbanization on Sao Paulo's climate suggest higher temperatures affect convective rainfall, which occurs when warm air rises, condenses to form clouds, and produces extreme rain.

Other concerns are the depletion of coral reefs and mangrove forests on Brazil's coastlines, and decreases in biodiversity.

Migration Trends

Migration from drought in northeastern Brazil to cities has increased urban populations, putting more people at risk of displacement from flooding and landslides. Migration from the northeast is a historical trend in Brazil, as economic migrants have sought seasonal jobs in more productive agricultural regions, or moved permanently to southern cities. Projected declines in rainfall have led some to predict further increases in migration in northeastern Brazil, as the IPCC has reported. However, remittances from family members who leave Brazil's northeast support resilience for those who remain and may help to reduce migration.

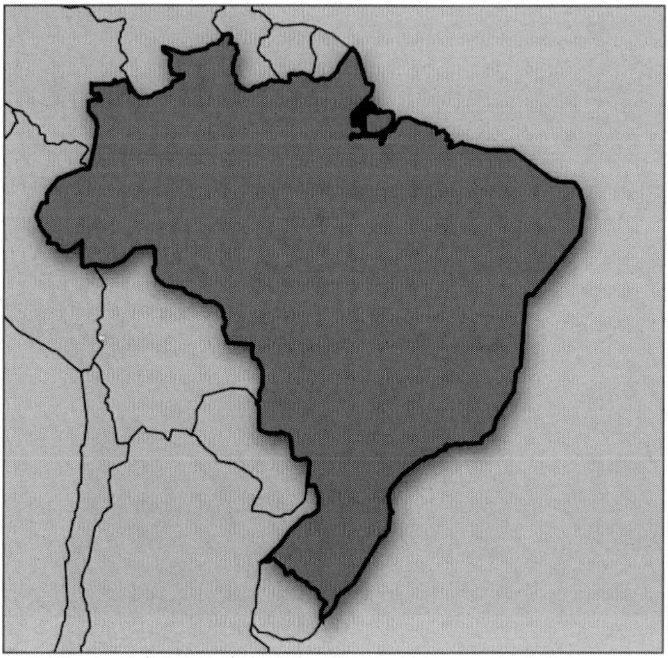

Source: Map Resources. | GAO-19-166.

Map of Brazil.

Already environmental factors contribute to migration to cities, including to favelas, informal settlements often constructed in hilly areas and floodplains outside of Brazilian cities. A significant number of the favela residents in Rio de Janeiro are migrants from northeastern Brazil, according to IOM. These new migrants may be at risk of further displacement if heavy rainfall, flooding, and other climate change effects destroy their vulnerable homes. For example, heavy rainfall in April 2010 resulted in landslides across Rio de Janeiro, displacing an estimated 5,000 people, according to a report from the World Bank. Brazil is also a destination for migrants from other countries in the region. Migrants from Venezuela searching for jobs and improved food security have come in growing numbers in recent years, as have migrants from Haiti fleeing a series of natural disasters, as IOM has reported.

Challenges in Stability and Security

Although Brazil ranks 106th out of 178 countries on the Fragile States Index, the effects of climate change may contribute to challenges with water, food, and energy access according to the IPCC. Decreased rainfall could decrease agricultural productivity, with potential health impacts for poor populations.

Source: World bank/Scott Wallace. | GAO-19-166.

A favela on the outskirts of Salvador de Bahia, Brazil.

These conditions are of particular concern in northeastern Brazil, as extreme weather and low crop yields are associated with more violence, according to the IPCC. Brazil also receives about 70 percent of its electricity from hydroelectric power, according the United Nations Environment Programme, and recent droughts caused power cuts across many major cities. Although not linked to the effects of climate change, absorbing a growing number of migrants fleeing political and economic instability in Venezuela may impact the broader region, according to the U.S. Department of Defense and the National Intelligence Council. Neighboring countries, including Brazil, may struggle to absorb the influx of migrants. On average, 800 Venezuelans are crossing the border to Brazil every day in need of urgent humanitarian assistance, according to the UNHCR, the UN Refugee Agency.

The Arctic

The effects of climate change in the Arctic, including higher temperatures and melting ice, have contributed to shifts in migration across the Arctic, and may have security implications. Increasing temperatures may have a variety of impacts in the Arctic, according to the Intergovernmental Panel on Climate Change (IPCC).

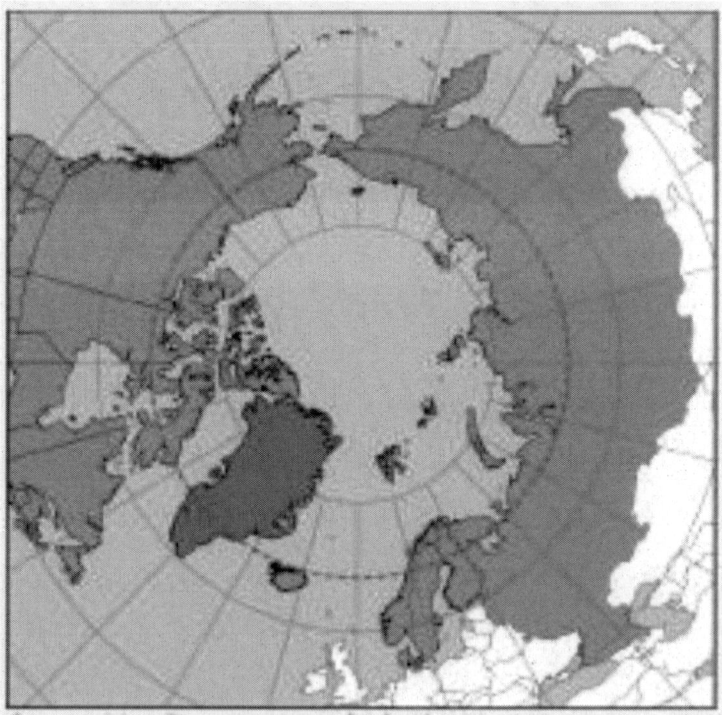

Source: Map Resources. | GAO-19-166.

The effects of rising temperatures are disrupting livelihoods and food security, especially for indigenous communities, and opening up untapped natural resources to extraction. Both trends have impacted migration flows in the Arctic. Rising temperatures and melting ice have opened up previously inaccessible waterways in the Arctic, with implications for national security, according to the Department of Defense and others. Greenland, located in the Arctic and considered part of Kingdom of Denmark, exhibits many of these trends.

Greenland: Climate Change, Migration, Stability and Security

Total Population	56,000
Fragile States Index	Not Available*
Human Development Index	Not Available*
Gross Domestic Product (GDP) per Capita	$48,159.5
Remittances as % of GDP	Not Available*
Agriculture, Fishing, Forestry as % of GDP	18.7
% Population in Cities	Not Available*
Net Migration Rate per 1000 people	Not Available*

* Statistics available for the Kingdom of Denmark, but not for Greenland, a part of Denmark.

Observed and Projected Effects of Climate Change

Greenland is experiencing the effects of climate change, including glacial and ice melt, shifts in wildlife distribution, and newly available oil and mineral deposits, among others. The Greenland Ice Sheet covers approximately 80 percent of Greenland's land mass. The ice sheet's melting rate is slow, but uncertain. Increases in temperature greater than 1°C may result in the near loss of the entire ice sheet over a millennium and significant sea level rise, according to the IPCC. In the short term, predicting the ice sheet's melting rate is a challenge as predictions vary in the scientific community. Accurate predictions would support mitigation and adaptation efforts in vulnerable areas. Rising temperatures and shrinking ice cover have shifted the distribution and migration patterns of marine mammals and fish, and impacted food security according to the IPCC and the Arctic Council, an intergovernmental forum for Arctic states. For example, the economy in Paamiut, Greenland, depended primarily on cod fisheries until changing climate conditions caused cod to disappear, and the town was slow to adapt to newly available shrimp. Similarly, fisheries in Disko Bay, Greenland, have struggled to adapt to new conditions. Rising temperatures and the resulting reduction in ice cover have required a shift to fishing from boats in open water instead of hunting and fishing over ice cover. Lastly, warming and ice melt may make significant oil and mineral deposits accessible for extraction in the future. The potential expansion of extraction industries makes environmental sustainability another possible concern. For example, an estimated 31 billion barrels of oil and gas may exist off the coast of Northeast Greenland, according to the Kingdom of Denmark's 2011-2020 Arctic Strategy. The strategy stresses the importance of assessing and reducing risks to the environment resulting from the exploration and extraction of oil and gas.

Migration Trends

The effects of climate change are predicted to contribute to internal and external migration in Greenland. For example, young people are increasingly leaving indigenous communities in rural areas for cities in Greenland in search of work, as traditional livelihoods become unsustainable. Greenland is home to a majority indigenous population, primarily Inuit, whose traditional hunting and fishing practices require travel across ice. In the past, people adapted to seasonal changes to support livelihoods by migrating, and the practice was embedded into indigenous social structures. With reduced ice cover, however, migrating to hunt, fish, and maintain connections to community is more dangerous or restricted. Government policies promoting centralized services, such as health care and education, have also played a role in the shift away from migration as a way of life. As a result, indigenous livelihoods are more difficult to maintain, and young people often migrate to towns and cities in Greenland, or to Denmark, for education. At the same time, warmer temperatures have made mineral extraction feasible. As the extraction industry grows, new jobs may draw migrants from outside the Arctic region. In 2011 companies spent $100 million on the exploration of minerals in the Artic, and the estimated number of new mines is expected to require more workers than now live in the region.[81] Currently, more people leave than migrate to Greenland.

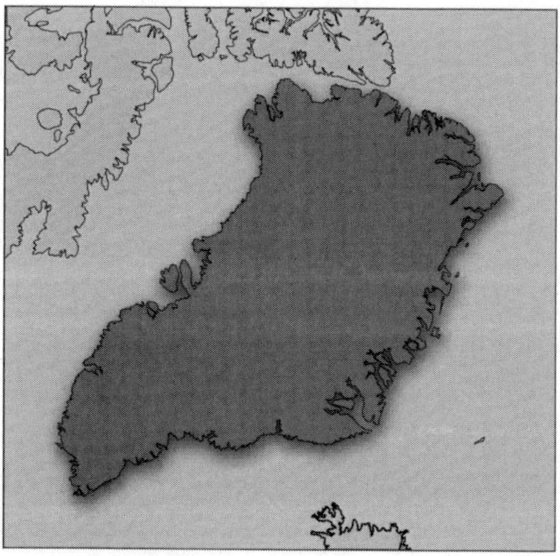

Source: Map Resources. | GAO-19-166.

Map of Greenland.

[81] Brookings-LSE Project on Internal Displacement, A Complex Constellation: Displacement, Climate Change and Arctic Peoples (January 30, 2013).

Challenges in Stability and Security

Although instability is not a concern in Greenland, the effects of climate change have increased its strategic importance to Denmark and other countries, which may impact indigenous society and governance. Denmark and its territories rank very high on the human development index, and the country is not considered a fragile state. However, activity in the Arctic has increased demand for defense capabilities in Greenland.

Source: GRID-Arendal/Lawrence Hislop. | GAO-19-166.

The local Inuit population in Uummannaq, Greenland relies heavily on ice coverage for fishing and travel by traditional dog-sled.

The retreat of sea ice in the Arctic, as reported by the U.S. National Snow and Ice Data Center, combined with an expected growth in human activity, has heightened U.S. interests in the Arctic region, including in Greenland, as we have previously reported.[82] In recent years Denmark has invested in defense capabilities specific to the Arctic, citing the need for

[82] GAO, *Coast Guard: Arctic Strategy Is Underway, but Agency Could Better Assess How Its Actions Mitigate Known Arctic Capability Gaps*, GAO-16-453 (Washington, D.C.: June 15, 2016).

improved operational capacity in the region. Disruption to indigenous ways of life, specifically hunting and fishing livelihoods, is also a concern. For example, high rates of alcoholism and domestic violence in Greenland have been linked to rapid social changes, according to a recent report from the European Union. Although a territory of Denmark, Greenland has considerable independence and may become more economically independent as revenues from the extraction of natural resources grow.[83]

SUB-SAHARAN AFRICA

The effects of climate change on Sub-Saharan Africa vary depending on the region and have impacts on migration and security, according to the International Organization for Migration (IOM). Coastal areas, for example, in West and East Africa are at risk from sea level rise that could affect major cities. Drought and the risk of desertification in the Sahel is cited as a concern, as is increased rainfall in parts of Central Africa accompanied by lower agricultural yields. As desertification threatens the livelihoods of farmers and herders, and drought makes fishing more challenging, rural dwellers may be more likely to migrate to cities, according to the United Nations Environment Programme (UNEP).

Urbanization and population growth across Sub-Saharan Africa is already making densely populated cities vulnerable to flooding, storms, and erosion, increasing the number of people at risk of displacement by sudden-onset disasters. Climate change effects and changing migration flows across Sub-Saharan Africa may impact access to natural resources and contribute to existing tensions and conflicts, according to UNEP and the Intergovernmental Panel on Climate Change (IPCC). In Nigeria, the effects of climate change may effect a variety of livelihoods and increase migration south, while also exacerbating existing conflicts.

---

[83] Brookings-LSE Project on Internal Displacement.

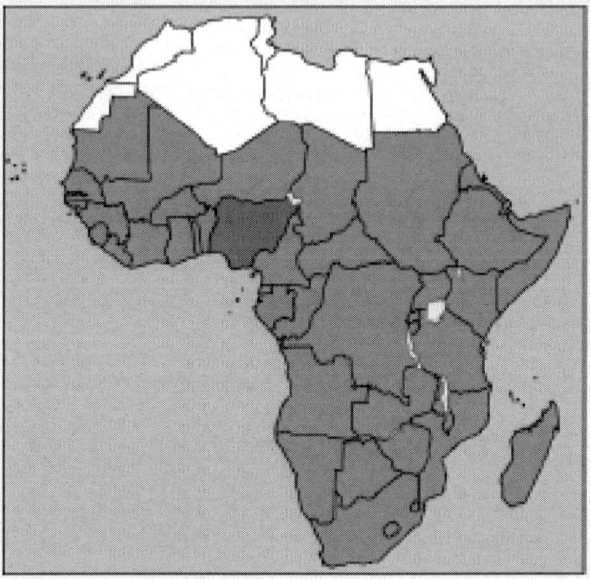

Source: Map Resources. | GAO-19-166.

Nigeria: Climate Change, Migration, Stability and Security

Total Population
190,900,000
Fragile States Index
#14 out of 178
A lower number indicates greater instability. Human Development Index Low
Gross Domestic Product (GDP)
per Capita
$5,338
Remittances as % of GDP
5.8
Agriculture, Fishing, Forestry as % of GDP
20.8
% Population in Cities
49.5
Net Migration Rate per 1000 people
-0.4

Observed and Projected Effects of Climate Change

The effects of climate change on Nigeria may impact the country's agriculture and economy, according to the United States Institute of Peace (USIP). Higher temperatures and decreased rainfall have contributed to drought in northern Nigeria. Desertification is also a concern. Some regions in northern Nigeria have less than 10 inches of rain a year, an amount that has decreased by 25 percent since the 1980's, according to USIP. In other areas across Nigeria flooding has resulted in major crop losses, according to UNEP. Rising sea level, water inundation, and erosion are concerns in Nigeria's coastal areas. Rising sea level is predicted to pose medium to very high risks to Africa's coastal areas by 2100, according to the IPCC. Future sea level rise could result in the inundation of over 70 percent of the Nigerian coast. A rise of 0.2 meters in sea level could risk billions of dollars in assets, including oil wells near the coast. Even without a rapid rise in sea level, Nigeria's coastal areas could experience erosion and significant land loss by 2100, as the IPCC has reported.

Migration Trends

The effects of climate change on livelihoods in northern Nigeria may contribute to migration to the south according to UNEP, while conflict in the north drives separate migration trends. As the effects of climate change make farming and fishing more challenging elsewhere in Nigeria, migration to southern coastal cities may increase. Traditionally, farmers, herders, and fishery workers migrated for temporary employment during the off season, including migration to Nigeria's cities to work in the oil industry. Permanent migration south as well as to cities may become more common if land suitable for farming decreases. As fish habitats like Lake Chad dry up, fishery workers may also migrate. Larger urban populations on the coast will put more people at risk of sea level rise, water inundation, and erosion, according to the IPCC. A rise in sea level of 1 meter could put over 3 million people at risk of displacement as the IPCC has reported. Herders have also moved further south due to increased drought in northern Nigeria, as UNEP and USIP have reported. A 2010 survey of herdsmen in Nigeria, for example, found that nearly one-third of them had migrated southeast as a result of changes in the natural environment, according to the UNEP. The ongoing conflict with Boko Haram, while not caused by climate change, has further resulted in millions of displaced people across the Lake Chad region, including many Nigerians who have fled to Cameroon, Chad, and Niger.

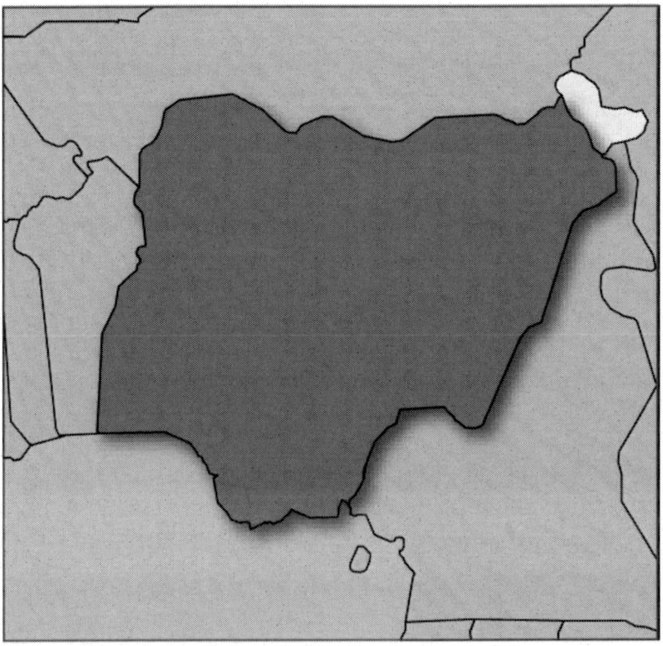

Source: Map Resources. | GAO-19-166.

Map of Nigeria.

Challenges in Stability and Security

The effects of climate change, migration, and conflict are interconnected in Nigeria, as USIP has reported. The country is ranked 14th of 178 countries on the Fragile States Index. Events in northwest Africa, including Boko Haram's attacks in Nigeria, have underscored concerns about the region's vulnerability to the spread of violent extremism.[84] The effects of climate change may exacerbate these concerns,

[84] GAO, *Combating Terrorism: U.S. Efforts In Northwest Africa Would Be Strengthened By Enhanced Program Management*, GAO-14-518 (Washington, D.C., June 24, 2014).

according to USIP. Nigerians fleeing attacks from Boko Haram in the north have gone to communities in neighboring Chad, Cameroon, and Niger that are already experiencing food shortages due in part to climate change. These neighboring countries as a result have fewer resources to support both their own residents and the newer refugees. Non-state actors may also take advantage of government inaction on the effects of climate change. Boko Haram, for example, has justified its acts of violence by pointing to government failures, according to the USIP. Separately, increased drought in the north may aggravate historic tensions over land and water use between farmers in the south and herders migrating from the north, according to UNEP. Nigeria's oil fields on the coast, which represent a significant part of the economy, are also at risk from sea level rise. Potential losses in oil revenue could impact Nigeria's ability to respond to humanitarian crises and conflict at home. Increased violence within its borders could also affect Nigeria's ability to support regional peacekeeping missions, such as the United Nations Mission in Liberia from 2003 to 2018, where Nigerian troops worked to restore security after a civil war.

Source: © UNHCR/D. Mbaiorem. | GAO-19-166.

MIDDLE EAST AND NORTH AFRICA

The effects of climate change in the Middle East and North Africa, including on its desert regions, may impact water access and compound migration and stability challenges, according to the United Nations Environmental Programme (UNEP). Over 60 percent of the population already experiences high or very high water stress, according to the World Bank. Coupled with unsustainable water use, climate change may further exacerbate challenges with water security. The region continues to experience rising temperatures and declining annual rainfall, trends that contribute to the severity and length of drought, land degradation, and desertification.

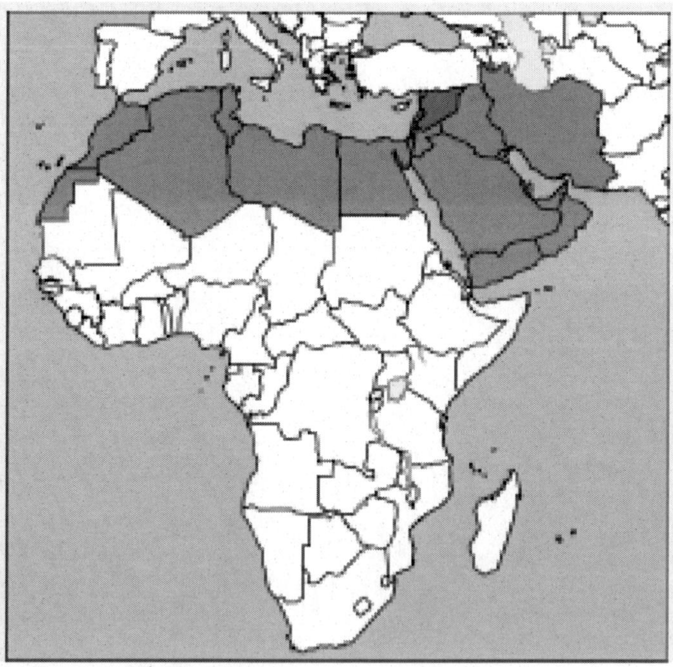

Source: Map Resources. | GAO-19-166.

Decreased water security affects the livelihood and quality of life of farmers in the region, contributing to an increase in their migration to the cities and more urbanization, according to the World Bank. In contrast, many people are expected to migrate away from coastal cities as a result of sea level rise, according to UNEP. These potential migrations would be taking place in a region that already hosts large numbers of migrants such as those displaced by conflict and violence, including 18 percent of the world's refugees, according to the International Organization for Migration. Challenges in water security may put greater pressure on unstable governments in the region, by intensifying existing tensions and conflicts between populations and their governments as well as between countries that share sources of water. The conflict in Syria illustrates the complex nature of climate change, migration, and conflict in the region, and the challenges to accurately assessing the links among the three, as noted in a technical paper commissioned by the U.S. Agency for International Development (USAID).

Syria: Climate Change, Migration, Stability and Security

Total Population
18,300,000
Fragile States Index
#4 out of 178
A lower number indicates greater instability.
Human Development Index
Low
Gross Domestic Product (GDP)
per Capita
Not available
Remittances as % of GDP
Not available
Agriculture, Fishing, Forestry as % of GDP
Not available
% Population in Cities
53.5
Net Migration Rate per 1000 people
-41.8

Rising temperatures and declining rainfall have contributed to recent droughts in Syria, a trend that may continue. The country underwent an extended drought from about 2006 until 2011. During the drought an estimated 60 percent of Syria experienced severe crop failure, and accompanying impacts on food security. Some studies have linked the length and severity of the drought in Syria to climate change, as USAID has reported. Others, however, have pointed to government land and water use policies, combined with the effects of climate change, as responsible for the severity of the drought. Agricultural policies, for example, encouraged farmers to grow water intensive crops like wheat, and supported inefficient irrigation practices, policies which further depleted ground water and made the region more vulnerable to decreases in rainfall linked to climate change. Across the Middle East, the rising temperatures and declining rainfalls of recent decades may worsen, according to the World Bank. If these trends continue, countries in the Middle East, including Syria, could continue to experience periods of severe drought and reduced crop yields.

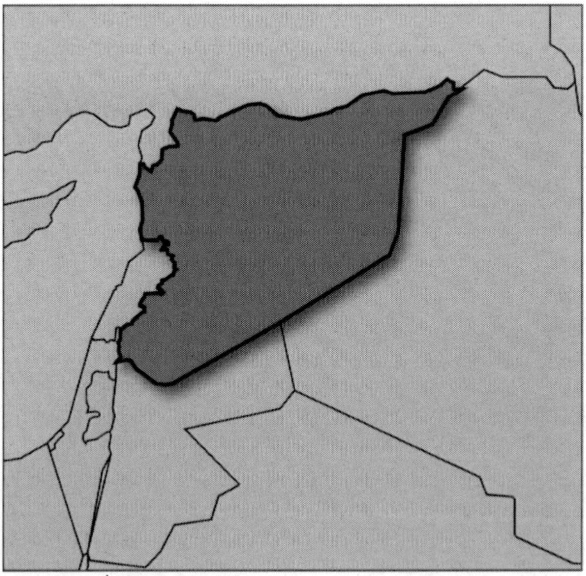

Source: Map Resources. | GAO-19-166.

Map of Syria.

Migration Trends

The ongoing conflict in Syria, in which migration due to climate change may have been a contributing factor, has caused large-scale migration to neighboring countries in the Middle East and to Europe. Leading up to the civil war, prolonged drought, among other factors, had increased migration to Syrian cities. Because of the drought, in 2009, over 800,000 Syrians lost their livelihoods in the agricultural sector, while nearly 1 million experienced food insecurity.

Source: © UNHCR/A. McDonnel. | GAO-119-166.

Za'atri camp in northern Jordan.

In 2010, an estimated 200,000 people migrated from farms in rural areas to cities, according to a UN report. The conflict in Syria, which began in 2011, has further displaced large numbers of people within the country and across the Middle East, as we have previously reported.[85] At the beginning of the conflict, Syrians, as well as Iraqi and Palestinian refugees who had been residing in Syria, fled mainly to Jordan, Lebanon, and

[85] GAO, *Syrian Refugees: U.S. Agencies Conduct Financial Oversight Activities for Humanitarian Assistance but Should Strengthen Monitoring*, GAO-18-58 (Washington, D.C.: Oct. 31, 2017).

Turkey. As the conflict persisted, refugees fled in larger numbers to Turkey, with the UNHCR reporting that nearly 1 million Syrians sought protection in that country in 2015. Starting that year, a growing number of Syrians risked dangerous sea voyages to reach countries in Europe, such as Greece, Germany, and Sweden. As of June 2017, more than 5 million registered Syrian refugees were living in neighboring countries, including more than 3 million in Turkey, and more than 1 million in Lebanon.

Challenges in Stability and Security

Sources agree that the Syrian conflict is a significant security challenge that has resulted in large scale migration across the Middle East and to Europe. Yet the link between prolonged drought, rural to urban migration, and the current conflict in Syria is uncertain. Some academic sources argue that the increased strain on urban infrastructure and resources due to the rural to urban migration played a role in Syria's growing instability. Others highlight the complex nature of the Syrian conflict, pointing to broader political factors that exacerbated resource scarcity and inequality. For example, as the drought intensified, the Syrian government downplayed the severity of the humanitarian crisis, as described in research cited in a technical report commissioned by USAID.[86] As a result, appeals to the international community for emergency aid received minimal support. Combined with existing sectarian divisions, ongoing revolutions across the Middle East, and other factors, the government's response to the drought may have contributed to the current conflict. Migration and displacement are a concern in the region, according to the Department of Defense and others. The U.S. government has provided significant humanitarian assistance for Syrian refugees in the Middle East, including in Lebanon and Jordan, as we have previously reported.[87]

---

[86] Schuyler Null and Lauren Herzer Risi, Woodrow Wilson International Center for Scholars, USAID Office of Conflict Management and Mitigation Discussion Paper, *Navigating Complexity: Climate, Migration, and Conflict in a Changing World* (November 2016).

[87] GAO-18-58.

However, a technical report commissioned by USAID has cautioned that the ongoing conflict in Syria makes it difficult to conduct research and draw conclusions related to climate, migration and conflict.[88]

OCEANIA

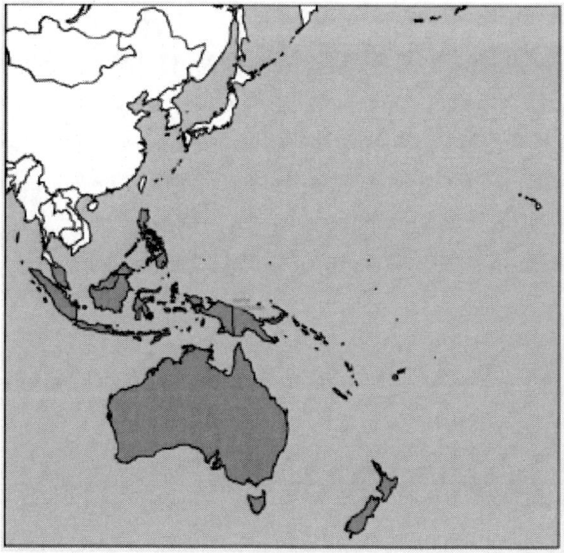

Source: Map Resources. | GAO-19-166.

The effects of climate change on Oceania, particularly rising seas, may significantly impact coastal populations and increase migration in the future, as the Asian Development Bank (ADB) and the Intergovernmental Panel on Climate Change (IPCC) have reported. Rising temperatures and declining rainfall may also contribute to lower yields from fisheries and agriculture, and a significant decrease in coral reef cover. Extreme weather events, including higher temperatures, wind, and rainfall, have already increased in number and intensity across the region. In the majority of Pacific island nations, of those who migrate, more people leave than come, according to the African, Caribbean, and Pacific Observatory on

[88] Null and Risi.

Migration. The majority of migration in the region is economically driven. In the future, climate change may further impact these migration patterns across the region, according to the IPCC.

Climate change has already exacerbated challenges that aid-dependent nations in the region face, restricting livelihoods and resources and contributing to pressures to migrate. The costs of climate change, including a decline in crop yields, a rise in energy demands, and a loss of coastal land, are predicted to be significant. The ADB estimates these costs will reach 12.7 percent of the Pacific regions' GDP by 2100. Increased migration may also impact political stability and play a role in geopolitical rivalries within the region, according to the IPCC. The effects of climate change, especially rising sea levels, may result in forced migration from the Republic of the Marshall Islands (the Marshall Islands) and have additional impacts on the U.S. defense infrastructure on the islands.

Marshall Islands: Climate Change, Migration, Stability and Security

Total Population	100,000
Fragile States Index	Not Available
Human Development Index	High
Gross Domestic Product (GDP) per Capita	$3,819
Remittances as % of GDP	14.8
Agriculture, Fishing, Forestry as % of GDP	15.9
% Population in Cities	76.7
Net Migration Rate per 1000 people	Not Available

Observed and Projected Effects of Climate Change

Rising sea levels are a grave threat to the Marshall Islands. The country consists of islands, low-lying atolls—coral caps sitting on top of submerged volcanoes—making it particularly vulnerable to rising sea levels.

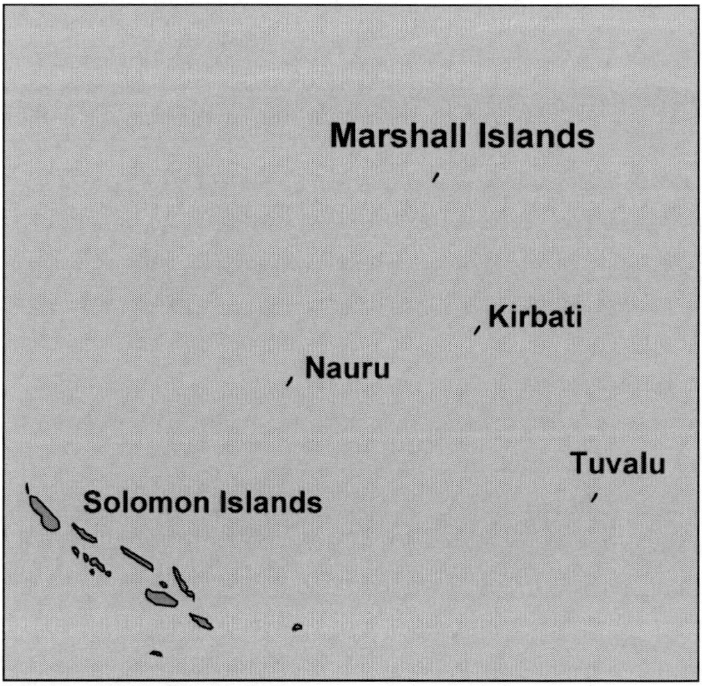

Source: Map Resources. | GAO-19-166.

Map of Marshall Islands.

On average, the Marshall Islands are 2 meters above sea level. In Majuro, the country's most populous atoll, observed rates of sea level rise are already twice as fast as the global average. Population centers experience significant flooding, with damage to roads, houses, and infrastructure, especially during La Niña years, which are significantly wetter and more prone to extreme rainfall. Flooding is expected to worsen with rising sea levels, with consequences for the availability of drinking water. On Roi-Namur island, for example, a 0.4 meter rise in sea level combined with wave-driven flooding is predicted to make groundwater undrinkable year round as early as 2055. This salt water inundation may contaminate already limited groundwater across the Marshall Islands. Lastly, during the 1940s and 1950s, the Marshall Islands was the site of 67 U.S. nuclear weapons tests on or near Bikini and Enewetak Atolls. Projected increases in frequency of flooding may negatively impact efforts to contain radioactive material stored on Runit Island.[89]

Migration Trends

A number of factors have increased migration from the Marshall Islands, including to the United States. In 1986, the United States entered into a compact of free association with the country that allowed its citizens to migrate to the United States, as we have previously reported. As a result, more than 20,000 Marshallese now live in the United States.[90] People are more likely to migrate abroad as the effects of climate change on the Marshall Islands—including rising sea levels—increasingly impact livelihoods.[91] The threat of mass displacement and forced migration is also a concern, as the International Organization for Migration has reported. However, Marshallese culture has a strong connection to the land, which means that many view migration as a last resort. For people still living in the Marshall Islands, they face overpopulation in urban centers and displacement by sudden-onset disasters like cyclones and flooding. Factors influencing people deciding to move abroad include displacement, lack of economic opportunity—sometimes exacerbated by climate change—and limited access to health care. Climate change is likely to increase risks to public health in the country[92] Increased rainfall, for instance, may expand

---

[89] Storlazzi, C.D., et al., 2017, *The Impact of Sea-Level Rise and Climate Change on Department of Defense Installations on Atolls in the Pacific Ocean (RC-2334): U.S. Geological Survey Administrative Report for the U.S. Department of Defense Strategic Environmental Research and Development Program.*

[90] GAO has previously reported on migration from the Marshall Islands. See GAO, Compacts of Free Association: Improvements Needed to Assess and Address Growing Migration, GAO-12-64 (Washington, D.C., November 14, 2011).

[91] International Organization for Migration, Republic of the Marshall Islands IOM Country Strategy 2017-2020 (Delap, Majuro, Republic of the Marshall Islands: 2017).

[92] Republic of the Marshall Islands, Joint National Action Plan for Climate Change Adaptation and Disaster Risk Management 2014-2018.

mosquito breeding grounds, raising the risk of diseases like dengue fever. The country's limited health care system may further contribute to migration from the islands.

Challenges in Stability and Security

In the future, the Marshall Islands may become uninhabitable. This prospect threatens the existence of the Marshall Islands as a sovereign state, as well as the United States defense facilities located on the islands. The total loss of land could result in the Marshall Islands being uninhabitable, which raises problems of migration, resettlement, cultural survival, and sovereignty. Relocation of the population of the Marshall Islands, and of other Pacific Island nations at risk of rising seas, could cause significant geopolitical challenges.[93]

Source: U.S. Geological Survey. | GAO-19-166.

Wave-driven flooding and overwash on Roi-Namur Atoll, Republic of the Marshall Islands.

[93] Storlazzi, C.D., et al., 2017.

The Marshall Islands are also of strategic importance for the United States. Under the Compact of Free Association, the United States has permission to use several islands— including Kwajalein Atoll, the location of the Ronald Reagan Ballistic Missile Defense Test Range—until 2066. The country's proximity to the equator makes the Marshall Islands ideal for missile defense and space work. Yet the island's defense infrastructure and operations are at significant risk due to rising sea levels, flooding, and diminishing supplies of potable water. As the Department of Defense has noted, climate change will have serious implications for the department's ability to maintain its infrastructure and ensure military readiness in the future.[94]

Central America and the Caribbean

The effects of climate change on Central America and the Caribbean may increase migration and exacerbate poverty rates, as the National Intelligence Council has reported. The climate in Central America and the Caribbean is predicted to be warmer and dryer. The Caribbean's extensive coastlines and low-lying areas are vulnerable to sea level rise and an increase in sudden-onset disasters, including hurricanes and storm surges. Drought is a particular concern in Central America, where declines in rainfall have reduced crop yields and threatened livelihoods in recent years. Some evidence shows that drought in parts of Central America has contributed to migration north, including to the United States. Population growth, especially in coastal cities, has increased the number of people at risk during hurricane season, and the number and intensity of hurricanes have grown in recent years. Some attribute the increase in intensity to higher sea surface temperatures caused by climate change. However, there remains debate about long term hurricane trends.

---

[94] DOD, *2014 Climate Change Adaptation Roadmap* (Alexandria, VA: June 2014).

Recent hurricanes have caused displacement, and significant losses and damages—including to infrastructure—across the region. The depletion of coral reefs and mangrove trees, natural barriers to coastal erosion and flooding, has exacerbated vulnerability to storms in coastal areas. Climate change is likely to have negative impacts on tourism in the Caribbean, where the industry is an important part of the economy, according to Inter-American Development Bank. Climate change impacts on the economy may make it increasingly difficult for governments to reduce poverty and move towards environmental sustainability. Haiti's geography, location, and high poverty rates make the country especially vulnerable.

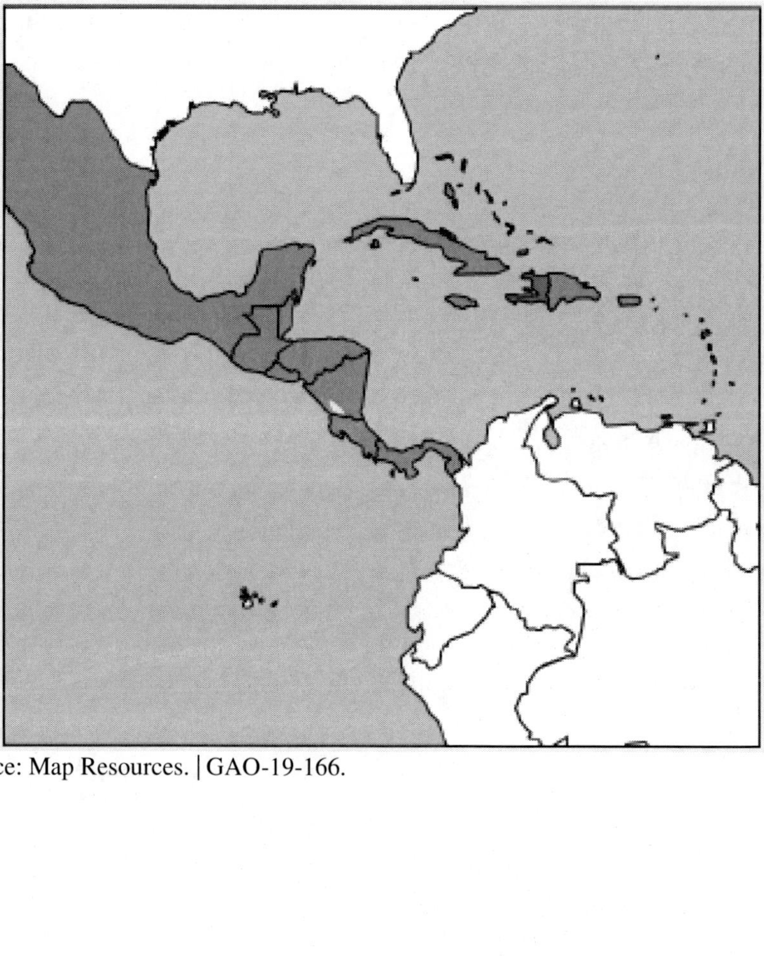

Source: Map Resources. | GAO-19-166.

Haiti: Climate Change, Migration, Stability and Security

Total Population
11,000,000
Fragile States Index
#12 out of 178
A lower number indicates greater instability.
Human Development Index
Low
Gross Domestic Product
(GDP) per Capita
$1,653
Remittances as % of GDP
29.3
Agriculture, Fishing, Forestry as % of GDP
17.6
% Population in Cities
54.3
Net Migration Rate per 1000 people
-2.9

Observed and Projected Effects of Climate Change

Haiti is highly vulnerable to climate change effects, partly due to its long coastline.[95] Hurricanes routinely make landfall in the country, and increases in rainfall and wind speeds associated with hurricanes are likely.

Severe hurricanes, including Hurricane Matthew in September 2016, have hit Haiti in recent years. Hurricane Matthew was the first category 4 storm in Haiti since 1964. Damage from severe flooding and severe winds during the hurricane affected over 2 million people and created significant food security and public health challenges. Significant deforestation has further exacerbated Haiti's vulnerability to hurricanes, as trees previously provided a natural barrier to the erosion that strong winds and more rainfall can cause. Rising temperature and highly variable rainfall have led to extreme drought and flash flooding, according to the U.S. Agency for

[95] International Organization for Migration, *The Atlas of Environmental Migration* (New York: Routledge, 2017).

International Development (USAID).[96] These trends decrease crop yields, affecting the livelihoods of farmers, and threaten water access. Projected increase in temperature and decreases in rainfall are likely to intensify drought in Haiti's interior.

Migration Trends

Slow-onset climate events, such as drought, and rising sea levels, and sudden-onset events, including earthquakes, affect Haiti, according to the International Organization for Migration (IOM). Haiti is also particularly exposed to extreme weather events, such as hurricanes, which can lead to displacement. In January 2010, a catastrophic earthquake in Haiti killed an estimated 230,000 people and left close to 1.5 million people homeless. According to IOM, the recurrence of environmental disruptions increases risks and vulnerabilities. When Hurricane Sandy struck Haiti in October 2012, the country had still not recovered from the 2010 earthquake. The worsening of climate change effects around the world, particularly in low-income countries, may increase the number of people wanting to immigrate to the United States, where approximately 700,000 Haitians live today.[97] Remittances from family members living outside Haiti make up a significant portion of the economy, at 24.7 percent of GDP. The majority of these remittances come from the United States, as we have previously reported.[98] Remittances may support resilience to climate change effects as migrants send money home for disaster recovery and adaptation.[99]

---

[96] USAID, *Haiti: Environment and Climate Change Fact Sheet* (January 2016).

[97] The Department of Homeland Security (DHS) designated Haiti for Temporary Protected Status (TPS) in the aftermath of the 2010 earthquake, allowing eligible Haitian nationals to apply for TPS. The TPS designation for Haiti was set to expire on January 22, 2018. On November 20, 2017, DHS determined that conditions in Haiti no longer supported its designation for TPS, and decided to terminate the designation for Haiti. Subsequently, in October 2018, a federal court enjoined DHS from implementing and enforcing the decision to terminate TPS for Haiti pending further resolution of the case, an order that DHS has appealed.

[98] GAO, *Remittances To Fragile Countries: Treasury Should Assess Risks from Shifts to Non-Banking Channels*, GAO-18-313 (Washington, D.C., March 8, 2018).

[99] Kanta Kumari Rigaud, Alex de Sherbinin, Bryan Jones, Jonas Bergmann, Viviane Clement, Kayly Ober, Jacob Schewe, Susana Adamo, Brent McCusker, Silke Heuser, and Amelia Midgley. 2018. *Groundswell: Preparing for Internal Climate Migration*. Washington, DC: The World Bank.

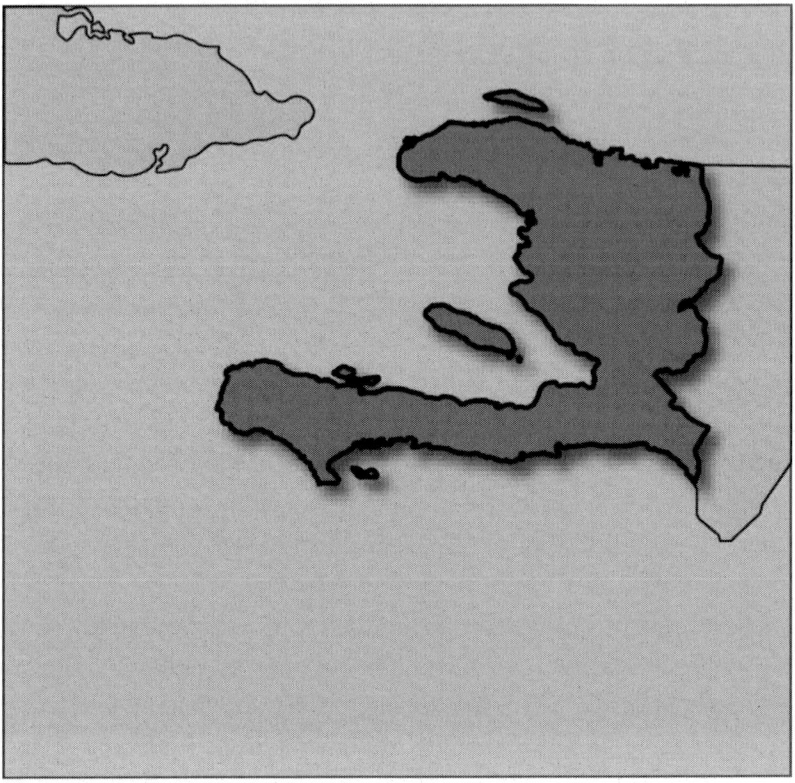

Source: Map Resources. | GAO-19-166.

Map of Haiti.

Challenges in Stability and Security

Haiti, the poorest country in the western hemisphere, has experienced political instability for most of its history, and ranks 12th of 178 on the Fragile States Index. The government has a low capacity to respond to additional challenges like those related to climate change, according to USAID. The Ministry of Environment, for example, is a relatively new organization within the Haitian government, and local and regional governments have a limited ability to enforce environmental laws and regulations. The United States has provided substantial aid to Haiti, both in disaster response and broader development projects. Official development assistance for Haiti in 2015, for instance, totaled slightly more than $1 billion. According to a January 2018 UN report, 2.8 million people were still in need of humanitarian assistance.

Source; UN Photo/Logan Abassi. | GAO-19-166.

Flooding in Cité Soleil, a neighborhood in Port-au-Prince, Haiti.

APPENDIX III: DEPARTMENT OF STATE GLOBAL CLIMATE CHANGE INITIATIVE ADAPTATION ACTIVITIES FUNDED IN FISCAL YEARS 2014 THROUGH 2017

The Department of State's Bureau of Oceans and International Environmental and Scientific Affairs (State/OES) provided about $78 million in adaptation funding from the Global Climate Change Initiative for eight projects for fiscal years 2014 through 2017 (see table 2).[100]

[100] The Global Climate Change Initiative was established in 2010 to promote resilient, low-emission development, and integrate climate change considerations into U.S. foreign assistance and was divided into three main programmatic initiatives: (1) Adaptation assistance, (2) Clean Energy assistance, and (3) Sustainable Landscapes assistance.

State/OES officials said that these projects aimed to help countries prepare for the impacts of climate change, potentially reducing the pressure to migrate, but to their knowledge, none of these projects directly supported activities related to migration. The president's budget for fiscal year 2018 did not include a request for funding for the Global Climate Change Initiative and State officials said that the agency does not plan to fund additional adaptation activities and has not requested additional funding for the activities.

Table 2. Department of State Global Climate Change Initiative Adaptation Activities Funded in Fiscal Years 2014 through 2017

Activity name	Countries in which activity occurred	Funding	Time period	Activity description
Adapt Asia-Pacific	Federated States of Micronesia, Fiji, Marshall Islands, Nauru, Papua New Guinea, Samoa, Solomon Islands, Tonga, Tuvalu, and Vanuatu	$2,000,000	September 2015 through November 2017	Implemented through the U.S. Agency for International Development, this project aimed to increase human and institutional capacity to mobilize and leverage international climate funds to improve adaptation and resilience in the Pacific region.
Increasing Private Investment in Landscape Restoration: A Pathway to Enhanced Resilience	Ethiopia, Niger, and Malawi	$197,530	August 2017 through March 2019	This project aims to enhance resilience by increasing countries' ability to attract investment in forest landscape restoration—a major pathway for achieving climate change resilience. The project will work with private sector investors, project developers, small- and medium-sized enterprises, governments, and communities to improve focal countries' ability to attract investment in adaptation.

Activity name	Countries in which activity occurred	Funding	Time period	Activity description
Least Developed Countries Fund (LDCF)	Global	$51,175,000	September 2015 through August 2017	The primary purpose of these contributions to the LDCF was to address the adaptation needs of the least developed countries, which are especially vulnerable to the adverse impacts of climate change. The LDCF financed the preparation and implementation of National Adaptation Programs of Action, which identify a country's priorities for adaptation actions.
National Adaptation Plans Global Network, Initial Grant	Albania, Belize, Benin, Botswana, Brazil, Cambodia, Colombia, Côte d'Ivoire, Dominica, Ethiopia, Fiji, Ghana, Grenada, Guinea, Guyana, Jamaica, Kenya, Kiribati, Liberia, Madagascar, Malawi, Mexico, Morocco, Nepal, Peru, Philippines, Samoa, Senegal, Sierra Leone, Saint Lucia, Saint Vincent and the Grenadines, South Africa, Suriname, Tanzania, Thailand, Togo, Uganda, and Vanuatu	$4,000,000	September 2015 through 2020	Initial grant to the National Adaptation Plans Global Network. The network is focused on increasing the capacity of national and subnational governments to identify and assess climate risks, integrate these risk considerations in sector planning, develop a pipeline of projects to address risks, identify and secure funding for projects, and track progress toward resilience targets.
National Adaptation Plans Global Network, U.S. in-country programs, Grant Cost Amendment	Colombia, East Caribbean (Guyana, Saint Lucia, Saint Vincent and the Grenadines), Ethiopia, Peru, South Africa, Uganda, West Africa (Côte d'Ivoire, Ghana, Guinea, Sierra Leone, Togo) and, under	$5,900,000	July 2016 through 2020	The cost amendment intensified the technical support on National Adaptation Plans to select countries dependent upon specific country adaption needs. In addition, the cost amendment continued the learning and progress from the initial grant.

Table 2. (Continued)

Activity name	Countries in which activity occurred	Funding	Time period	Activity description
	current consideration, East Caribbean (Dominica, Suriname), and Pacific (Fiji, Kiribati, Tuvalu)			
Pacific Catastrophe Risk Assessment and Financing Initiative	Cook Islands, Fiji, Kiribati, Marshall Islands, Federated States of Micronesia, Nauru, Niue, Palau, Papua New Guinea, Samoa, Solomon Islands, Tonga, Tuvalu, and Vanuatu	$8,000,000	September 2016 through September 2017	Implemented through the Department of Treasury, this funding supported a Treasury grant to the Pacific Catastrophe Risk Assessment and Financing Initiative Multi Donor Trust Fund at the World Bank. This activity established the Pacific Catastrophe Risk Insurance Foundation and the Pacific Catastrophe Risk Insurance Company, among other things.
Private Investment for Enhanced Resilience (PIER)	Bangladesh, Ghana, Guyana, Indonesia, Mozambique, Peru, Tanzania, and Vietnam	$5,382,715	September 2017 through September 2020	The goal of PIER is to increase private sector investment in resilience to climate change in eight developing countries. The first phase of the project will assess and identify opportunities for private investment in resilience, as well as build public and private capacity for climate risk assessment in all the countries. In the second phase, public and private sector partners will develop and pilot climate risk-reduction investment models in four of the countries. The third phase will publicize the piloted investment models and lessons learned among the eight countries.

Activity name	Countries in which activity occurred	Funding	Time period	Activity description
U.S.-India Partnership for Climate Resilience	India	$1,150,000	September 2015 through September 2019	Implemented through the National Oceanic and Atmospheric Administration, this activity aims to implement a capacity-building partnership with India to promote effective climate resilient decision making at national, state, and local levels.

Source: GAO analysis of Department of State project information from the Bureau of Oceans and International Environmental and Scientific Affairs. | GAO-19-166.

APPENDIX IV: COMMENTS FROM THE DEPARTMENT OF STATE

United States Department of State
Comptroller
Washington, DC 20520

December 21, 2018

Thomas Melito
Managing Director
International Affairs and Trade
Government Accountability Office
441 G Street, N.W.
Washington, D.C. 20548-0001

Dear Mr. Melito:

We appreciate the opportunity to review your draft report, "CLIMATE CHANGE: Activities of Selected Agencies to Address Potential Impact on Global Migration, GAO Job Code 102302.

The enclosed Department of State comments are provided for incorporation with this letter as an appendix to the final report.

Sincerely,

Christopher H. Flaggs

Enclosure:
As stated

cc: GAO – David Gootnick
F – Eric Ueland
OIG - Norman Brown

Department of State Comments on GAO Draft Report

CLIMATE CHANGE: Activities of Selected Agencies to Address Potential Impact on Global Migration (GAO-19-166, GAO Code 102302)

Thank you for the opportunity to comment on the GAO draft report *"Climate Change: Activities of Selected Agencies to Address Potential Impact on Global Migration."*

GAO Recommendation: The Secretary of State should ensure that the Director of the Office of U.S. Foreign Assistance Resources provides missions with guidance that clearly documents the department's process for climate change risk assessments for integrated country strategies.

The Department does not oppose this recommendation. By June 30, 2019 we will update our Integrated Country Strategy (ICS) guidance and will specifically note that missions have the option to provide additional information on a variety of topics such as Climate Resilience, in the form of Annexes to their strategy.

In parallel to the above, F and BP will begin working with stakeholders to consider whether to recommend that the Secretary ask the President to rescind Executive Order 13677 – Climate-Resilient International Development.

APPENDIX V: COMMENTS FROM THE U.S. AGENCY FOR INTERNATIONAL DEVELOPMENT

USAID
FROM THE AMERICAN PEOPLE

December 18, 2018

David Gootnick
Director, International Affairs and Trade
U.S. Government Accountability Office
441 G Street, N.W.
Washington, D.C. 20548

Re: CLIMATE CHANGE: Activities of Selected Agencies to Address Potential Impact on Global Migration (GAO-19-166)

Dear Mr. Gootnick:

I am pleased to provide the formal response of the U. S. Agency for International Development (USAID) to the draft report of the U. S. Government Accountability Office (GAO) entitled, *"CLIMATE CHANGE: Activities of Selected Agencies to Address Potential Impact on Global Migration"* (GAO-19-166).

USAID appreciates the draft report's recognition of our Agency's ongoing efforts to support governments, civil society and the private sector in our partner countries to build resilience and manage risk that, if left unaddressed, could lead to costly humanitarian disasters and related displacement or migration. As the report indicates, the relationship between migration decisions and climate and extreme weather events is complex, mediated by a host of economic, political, social, and demographic factors.

USAID works to build local and national capacity to address this potential area of risk through four types of programs or initiatives: (1) humanitarian assistance and response to disasters, for which USAID serves the U.S. Government's lead coordinator; (2) programs that reduce the risk of disasters, such as by supporting the development of early-warning systems and response capacity; (3) efforts to build resilience to climate-related shocks and stresses, such as through the *Pastoralist Areas Resilience Improvement through Market Expansion* activity in Ethiopia; and, (4) institutional requirements that protect taxpayer investments across USAID's development portfolio by assessing, addressing, and adaptively managing risks that may be anticipated to arise from weather- and climate-related events and conditions.

In addition, USAID's Transformation, which has proposed to create three new Bureaus (for Resilience and Food Security, Conflict-Prevention and Stabilization, and Humanitarian Affairs), under the leadership of a new Associate Administrator for Relief, Response, and Resilience, will organize and elevate our work on the resilience-to-relief continuum, enhancing our ability to anticipate future crises, and work with public and private actors in partner countries to put in place the conditions to avoid or manage them.

I am transmitting this letter and the enclosed USAID comments for inclusion in GAO's final report. Thank you or the opportunity to respond to the draft report, and for the courtesies extended by your staff while conducting this engagement. We appreciate the opportunity to participate in the complete and thorough evaluation of our risk-management and resilience-building activities, and how they affect global migration.

Sincerely,

Angelique M. Crumbly
Acting Assistant Administrator
Bureau for Management

COMMENTS BY THE UNITED STATES AGENCY FOR INTERNATIONAL DEVELOPMENT (USAID) ON THE U.S. GOVERNMENT ACCOUNTABILITY OFFICE (GAO) DRAFT REPORT – CLIMATE CHANGE: Activities of Selected Agencies to Address Potential Impact on Global Migration (GAO-19-166)

The U. S. Agency for International Development (USAID) would like to thank the U. S. Government Accountability Office (GAO) for the opportunity to respond to this draft report. We appreciate the extensive work of the GAO engagement team.

The report contains no recommendation for USAID's action.

We are grateful for the report's recognition of USAID's programs that address the complex relationships among migration, disasters, climate and extreme weather-related events. These include the following:

- USAID's role as the lead Federal coordinator for international disaster assistance. The proposed Bureau for Humanitarian Assistance would advance our goal of delivering food and non-food international disaster aid seamlessly, which would serve the U.S. Government's foreign-policy interests and the people in need of humanitarian assistance more effectively. Specifically, the new Bureau would strengthen USAID's role in disasters overseas by consolidating core capabilities—including technical, policy, and operational expertise—into one, seamless entity, well-positioned to achieve maximum efficiency, effectiveness, and accountability, as well as elevate and strengthen USAID's lead humanitarian voice in the U.S. Government interagency, and with all international partners.

- USAID's programs to support the ability of other countries to withstand severe weather-related events. These activities, known as Disaster-Risk Reduction (DRR), are aimed at preventing new risk, reducing existing risk, and managing residual risks, all of which contribute to strengthening resilience and helping nations on their Journey to Self-Reliance. DRR investments include early-warning systems to notify vulnerable populations of imminent disasters; training programs for first-responders; changes to policies on disaster-response and risk-reduction at the national level; and strengthening shelters and other settlements to help withstand hazards, such as hurricanes or earthquakes.

- USAID's activities that support communities in building capacity and resilience to climate-related shocks and stresses. Examples include the *Pastoralist Areas Resilience Improvement through Market Expansion* activity in Ethiopia, which has been supporting pastoralist communities to manage the risk of drought, while building self-reliance, enhancing food security, and promoting private enterprise; as well as USAID's recent activity for *Climate-Resilient Ecosystems and Livelihoods* in Bangladesh; and an activity in Mali that strengthens the capacity of the country's national meteorological agency. We appreciate that the GAO highlighted these activities in the report.

- USAID's requirement, codified in Automated Directives Series Chapter 201, that nearly all new development strategies, projects, and activities assess climate-related risks, and that those with moderate or high levels of risk address and manage those risks adaptively. The objective of the management of climate risk is to ensure that development results and U.S. taxpayer investments are resilient to extreme weather and climate impacts. In Fiscal Year 2017, USAID assessed the climate risk of $6.3 billion of planned investments; as a result, many projects have incorporated approaches that will improve service-delivery, while enhancing sustainability and resilience.

In addition, we note that the proposed Transformation of USAID includes a proposal to elevate our work on building resilience under a new Bureau for Resilience and Food Security (BRFS), which would incorporate our climate-adaptation team. This new Bureau would better position USAID to integrate planning for risks, such as those highlighted in this report, into broader efforts to build resilience in our partner countries. The new BRFS would work with two other redesigned Bureaus —Conflict-Prevention and Stabilization, and Humanitarian Affairs — under the leadership of a new Associate Administrator for Relief, Response, and Resilience. This would position the Agency to elevate our work on the issues that arise along the resilience-to-relief continuum, and enhance our ability to anticipate future crises and work with governments, civil society and the private sector in partner countries to put in place the conditions to avoid or manage shocks. Complementing the climate-adaptation and resilience efforts of BRFS, other climate-related technical-leadership functions, including the management of climate-related risks, would reside in the proposed Bureau for Development, Democracy, and Innovation (DDI).

The answer to addressing and managing the complex set of factors that lead people to migrate is building safer, more resilient, and more prosperous societies, as well as responding effectively to disasters when they do occur. The programs described above contribute to the ability of USAID and the wider U.S. Government to do this work.

In: Climate Change
Editor: Jeremy Nesbitt

ISBN: 978-1-53617-216-4

Chapter 3

CLIMATE CHANGE: ANALYSIS OF REPORTED FEDERAL FUNDING*

United States Government Accountability Office

ABBREVIATIONS

ARM	Atmospheric Radiation Measurement
ASR	Atmospheric System Research
ATLAS	Advanced Topographic Laser Altimeter System
Commerce	Department of Commerce
CESD	Climate and Environmental Sciences Division
CESM	Climate and Earth System Modeling
CCTP	Climate Change Technology Program
CLARREO	Climate Absolute Radiance and Refractivity Observatory
CPF	Climate Absolute Radiance and Refractivity Observatory Pathfinder
DOD	the Department of Defense

* This is an edited, reformatted and augmented version of United States Government Accountability Office; Report to the Chairman, Committee on Science, Space, and Technology, House of Representatives, Publication No. GAO-18-223, dated April 30, 2018.

DOE	Department of Energy
ICESat	The Ice, Cloud, and Land Elevation Satellite
ICESat-2	The Ice, Cloud, and Land Elevation Satellite-2
LASP	Laboratory for Atmospheric and Space Physics
NASA	National Aeronautics and Space Administration
NSF	National Science Foundation
NOAA	National Oceanic and Atmospheric Administration
NIFA	National Institute of Food and Agriculture
NIST	National Institute for Standards and Technology
OMB	Office of Management and Budget
OCO-2	Orbiting Carbon Observatory-2
OCO-3	Orbiting Carbon Observatory-3
SORCE	Solar Radiation and Climate Experiment
TES	Terrestrial Ecosystem Science
TCSR	Terrestrial Carbon Sequestration Research
TSIS-1	Total Solar Irradiance Sensor-1
TSIS-2	Total Solar Irradiance Sensor-2
USGCRP	U.S. Global Change Research Program
USDA	U. S. Department of Agriculture
UV-B	ultraviolet radiation

Why GAO Did This Study

Since 1993, OMB has reported over $154 billion in funding for federal climate change activities, spread across the government—raising questions about fragmentation, overlap, or duplication.

GAO was asked to review federal climate change funding. This chapter examines (1) reported federal funding from 2010 to 2017 and the extent to which reports on such funding are clearly linked to the federal fiscal exposure to climate change; (2) the extent to which selected agencies reported climate change funding that supports programs where addressing climate change is the primary purpose; and (3) the extent to which the primary purpose programs are fragmented, overlapping, or duplicative.

GAO reviewed OMB climate change funding reports; analyzed budget justifications for six agencies—the Departments of Agriculture, Commerce, Defense, and Energy; the National Aeronautics and Space Administration and the National Science Foundation—representing 89 percent of OMB-reported climate change funding in fiscal year 2014; analyzed documents on primary purpose programs against GAO's fragmentation, overlap, or duplication criteria; and reviewed GAO's prior work on fiscal exposures.

What GAO Recommends

GAO is making two recommendations to OMB for enhancing the information it provides to Congress, in conjunction with future funding reports. OMB agreed with the findings but disagreed with GAO's recommendations, which GAO continues to believe are valid as discussed in the report.

What GAO Found

In its reports to Congress, the Office of Management and Budget (OMB) reported that annual federal climate change funding increased by \$4.4 billion from fiscal years 2010 through 2017. For example, reported annual funding for technology to reduce emissions increased by about \$3.5 billion, as seen in the Figure below. Although OMB included information on federal fiscal exposure to climate change in the President's budgets for fiscal year 2016 and 2017, it did not provide this information in its most recent climate change funding reports. For example, the reports did not include information on programs—such as disaster assistance— whose costs were likely to increase due to climate change which would have provided more complete information for making spending trade-off decisions for climate activities. According to GAO's prior work, more complete information on fiscal exposures and the long-term effects of decisions would help policymakers make trade-offs between spending with long-term and short-term benefits.

Based on its review of the budget justifications of six agencies representing 89 percent of OMB-reported funding, GAO identified few programs (18 of 533) whose primary purpose is to address climate change. The remaining programs were multi-purpose—the budget justifications included other program goals in addition to addressing climate change. The 18 programs represented about 6 percent of these agencies' reported climate change funding for fiscal year 2017.

According to GAO's analysis, the 18 primary purpose climate change programs GAO identified are fragmented across four federal agencies, but the programs serve different purposes, target different audiences, or operate at different time periods and scales, which minimizes potential overlap or duplication.

Additionally, agency program managers collaborate through the U.S. Global Change Research Program—a coordinating entity—to avoid potential negative effects from fragmentation. However, climate change programs outside GAO's review have not been analyzed for potential fragmentation, overlap, or duplication.

April 30, 2018

The Honorable Lamar Smith
Chairman
Committee on Science, Space, and Technology
House of Representatives

Dear Mr. Chairman:

Since 1993, the federal government has reported over $154 billion in funding for activities to understand and address climate change, according to data from the Office of Management and Budget (OMB). As reported by OMB, this funding has increased over time and is spread across the federal government. Specifically, federal climate change funding in nominal dollars increased from about $2.4 billion across 10 agencies in fiscal year

1993 to about $13.2 billion across 19 agencies in proposed budget authority for fiscal year 2017.[1] Recognizing that each department and agency operates under its own authorities and responsibilities—and can address climate change in different ways relevant to its own mission—federal efforts have followed a decentralized approach, with individual agencies deciding how to incorporate climate- related information into their planning, operations, policies, and programs. Entities within the Executive Office of the President, such as the Office of Science and Technology Policy and the U.S. Global Change Research Program (USGCRP), work together to coordinate federal climate change activities.

Given the current fiscal constraints facing the federal government, efforts to effectively use resources are particularly important. We previously reported on broad principles that could help policymakers make effective resource decisions during the budget process.[2] Specifically, we reported that a budget process should provide information about the long-term effect of decisions, provide information needed to understand important trade-offs, and provide for accountability and transparency. Additionally, in an October 2013 report, we reviewed risks facing the federal budget and found that for some risks, such as weather-related risks, the budget provided incomplete information or potentially misleading signals regarding the full cost of existing commitments.[3] In that report we concluded that, consistent with our prior recommendations to provide information on fiscal exposures in the federal budget process, developing

---

[1] When adjusted for inflation, reported climate change funding increased from $3.6 billion in 1993 to $12.9 billion in the proposed budget authority for 2017.

[2] GAO, *Fiscal Exposures: Improving Cost Recognition in the Federal Budget,* GAO-14-28 (Washington, D.C.: Oct. 29, 2013) and *The Nation's Fiscal Health: Action is Needed to Address the Federal Government's Fiscal Future*, GAO-17-237SP (Washington, D.C.: Jan. 17, 2017).

[3] GAO-14-28. According to the National Academies of Sciences, Engineering, and Medicine, the ability to attribute the causes of some extreme event types has advanced rapidly since the emergence of event attribution science a little more than a decade ago, while attribution of other event types remains challenging. Confidence in attribution of specific extreme events is highest for extreme heat and cold events, followed by hydrological drought and heavy precipitation. For example, for extreme heat and cold events in particular, changes in long-term mean conditions provide a basis for expecting that there also should be related changes in extreme conditions. National Academies of Sciences, Engineering, and Medicine. 2016. *Attribution of Extreme Weather Events in the Context of Climate Change*. Washington, D.C.: The National Academies Press. doi: 10.17226/21852.

additional information on these budgetary risks could help policymakers make better informed decisions.

In prior work, we also found that policymakers face challenges in making effective budget decisions for programs and activities to address climate change. Specifically, in May 2011, we reviewed federal climate change funding from fiscal years 2003 through 2010 and found that it was organized across the federal government in a complex, crosscutting system.[4] We also found that reported funding increased over time but was not aligned with strategic priorities because there was no coherent government-wide approach for addressing climate change. We recommended that the appropriate entities within the Executive Office of the President clearly establish strategic climate change priorities that consider the full range of federal efforts to address climate change.

Further, in November 2015, we found that the federal government's climate information—composed of climate data from satellites and weather stations, climate model projections, and other tools to inform decisions—was fragmented across many individual agencies and did not fully meet the needs of decision makers in federal, state, local, and private sectors.[5] We recommended that the Executive Office of the President direct a federal entity to develop a set of authoritative climate change projections and observations and create a national climate information system with defined roles for federal agencies and others. The Executive Office of the President neither agreed nor disagreed with the 2011 and 2015 recommendations and as of June 2017, had not implemented the recommendations.

The reported increase in funding and fragmentation of federal climate information activities raises questions about whether other federal climate change efforts, such as technologies to reduce emissions, may be fragmented, overlapping, or duplicative. This chapter uses definitions to describe fragmentation, overlap, and duplication among government

[4] GAO, *Climate Change: Improvements Needed to Clarify National Priorities and Better Align Them with Federal Funding Decisions*, GAO-11-317 (Washington, D.C.: May 20, 2011).

[5] GAO, *Climate Information: A National System Could Help Federal, State, Local, and Private Sector Decision Makers Use Climate Information*, GAO-16-37 (Washington, D.C.: Nov. 23, 2015).

programs from our prior work.[6] Fragmentation refers to circumstances in which more than one federal agency (or more than one organization within an agency) is involved in the same broad area of national need and opportunities exist to improve service delivery. Overlap occurs when multiple agencies or programs have similar goals, engage in similar activities or strategies to achieve their goals, or target similar beneficiaries.[7] Duplication occurs when two or more agencies or programs are engaged in the same activities or provide the same services to the same beneficiaries.

You asked us to review federal funding for climate change activities. This chapter examines (1) reported federal climate change funding since 2010 and the extent to which reports on such funding are clearly linked to the federal fiscal exposure to climate change; (2) the extent to which selected federal agencies reported climate change funding that supports programs for which addressing climate change is the primary purpose, versus programs for which addressing climate change is one of multiple purposes; and (3) the extent to which primary purpose climate change programs are fragmented, overlapping, or duplicative.

To address all three objectives, we (1) reviewed government-wide climate change funding reports by OMB from fiscal years 2010 through 2017, and (2) reviewed agency reports, budget justifications and planning documents for a non-generalizable sample of six departments or agencies—the Department of Commerce, the Department of Defense (DOD), the Department of Energy (DOE), the U.S. Department of Agriculture (USDA), the National Aeronautics and Space Administration (NASA), and the National Science Foundation (NSF). We selected these departments or agencies because they represented the largest share (89 percent) of OMB-reported climate change funding in 2014—the most recent OMB data available at the start of our review—and also conducted

---

[6] GAO, *Fragmentation, Overlap, and Duplication: An Evaluation and Management Guide*, GAO-15-49SP (Washington, D.C.: Apr. 14, 2015).

[7] We have reported that areas of overlap do not necessarily lead to duplication of efforts because initiatives sometimes differ in meaningful ways or leverage the efforts of other initiatives. See, for example, GAO, *Solar Energy: Federal Initiatives Overlap but Take Measures to Avoid Duplication*, GAO-12-843 (Washington, D.C.: Aug. 20, 2012).

activities in each of the climate change funding categories described in OMB's reports. The results of our analysis of these departments or agencies' activities cannot be generalized to those of departments or agencies not included in our review. We also interviewed officials from coordinating entities such as the Council on Environmental Quality, and USGCRP. We assessed the reliability of OMB and budget data used in our analysis by comparing data across multiple sources, reviewing documentation about the data, tracing data back to its source, and interviewing officials knowledgeable about the data. We determined these data were sufficiently reliable for the purposes of our reporting objectives and have noted in our report any limitations associated with the data we present.

To determine OMB-reported climate change funding since fiscal year 2010, we analyzed climate change funding data from OMB climate change funding reports by agency and line item. To assess whether these reports are clearly linked to federal fiscal exposure to climate change, we reviewed relevant reports by OMB and others that describe federal fiscal exposure to climate change impacts, and our prior work on providing information on federal fiscal exposures to policymakers to inform their decisions.

To determine the extent to which reported funding supports programs where addressing climate change is the primary purpose, we analyzed budget justifications for the six departments or agencies we selected against a list of climate change terms from OMB and USGCRP reports. For example, we reviewed agency program descriptions for references to "climate change," "renewable energy," "adaptation," and "climate resilience," among others. Based on the program purpose and activities described in agency budget justifications, we categorized programs as primary purpose or multiple-purpose. Primary purpose programs are those programs that agencies described as having addressing climate change as their main focus (e.g., a climate modeling program).

Multiple- purpose programs are those programs that agencies described as having several goals in addition to addressing climate change (e.g., a clean energy technology program designed to promote rural economic development).

To determine whether primary purpose programs were fragmented, overlapping, or duplicative, we reviewed additional program documentation—including authorizing legislation, strategic planning documents, and performance reports—and evaluated program characteristics based on our fragmentation, overlap, or duplication criteria. We also applied our fragmentation, overlap, or duplication criteria to additional multiple-purpose programs. Specifically, we selected one multiple-purpose program from each of the six selected departments or agencies and two programs from departments or agencies outside our review. We selected these programs because they either had the largest reported climate change funding amounts for a multiple-purpose program within that particular department or agency, or the program engaged in similar activities as primary purpose programs.

The term "funding" in this chapter refers to budget authority, or the authority provided by federal law to enter into financial obligations that will result in outlays involving federal government funds, as reported by OMB in its climate change funding reports required by law. The term "agencies" in this chapter refers to federal departments or agencies. The term "program" in this chapter refers to programs or activities that agencies identify as line items in their budget justifications. Unless otherwise stated, we report funding in nominal terms (not adjusted for inflation), and all years refer to fiscal years.

We conducted this performance audit from August 2016 to April 2018 in accordance with generally accepted government auditing standards.

Those standards require that we plan and perform the audit to obtain sufficient, appropriate evidence to provide a reasonable basis for our findings and conclusions based on our audit objectives. We believe that the evidence obtained provides a reasonable basis for our findings and conclusions based on our audit objectives.

Background

Since fiscal year 2010, annual appropriations laws have generally required the President to report on use of climate change funds. Specifically, the laws have generally required the President to report all federal agency funding, domestic and international, for climate change programs, projects, and activities to the congressional appropriations committees within 120 days of submitting the President's annual budget request.[8] In response to these requirements, OMB has issued three reports to congressional appropriations committees. Taken together, the three OMB reports describe climate change funding for fiscal years 2010 through 2017.[9] Generally, these reports included federal agency funding for the following three categories:[10]

[8] Annual appropriations acts prior to fiscal year 2010 have also often included a requirement for the President to issue a report on climate change funding. We have previously reported on the reports OMB issued in response to these prior requirements. In May 1999, we reported about OMB's April 1999 report summarizing federal agency programs related to global climate change. GAO, *Climate Change: Observations on the April 1999 Report on Climate Change Programs*, GAO-T-RCED-99-199 (Washington, D.C.: May 20, 1999). In August 2005, we reported on federal funding for climate change from 1993 through 2004 as reported by OMB. GAO, *Climate Change: Federal Reports on Climate Change Funding Should Be Clearer and More Complete*, GAO-05-461 (Washington, D.C.: Aug. 25, 2005). In May 2011, we reported on federal funding for climate change for fiscal years 2003 through 2010 as reported by OMB. GAO, *Climate Change: Improvements Needed to Clarify National Priorities and Better Align Them with Federal Funding Decisions*, GAO-11-317 (Washington, D.C.: May 20, 2011).

[9] Specifically, OMB issued reports in June 2010, August 2013, and January 2017. OMB did not issue some of the required reports, such as the report required by the Consolidated Appropriations Act for fiscal year 2017, and did not meet the reporting deadlines for the reports required by the Consolidated Appropriations Act for fiscal year 2014, the Consolidated and Further Continuing Appropriations Act for fiscal year 2015, and the Consolidated Appropriations Act for fiscal year 2016. OMB's report issued in January 2017 responded to the reporting requirements in these three laws.

[10] OMB has also reported funding for tax expenditures related to climate change, and in its August 2013 report, it included a Natural Resources Adaptation category that captured funding for adaptation activities as reported by the Department of the Interior. See appendixes III and V for more information on OMB-reported climate change funding by category since 1993, and appendix VI for more information on OMB-reported tax expenditures.

- Clean Energy Technology, which includes the research, development, and deployment of technologies to reduce greenhouse gas emissions and the reliance on fossil fuel, such as clean energy systems, geothermal, solar, wind, biomass, nuclear, and hydropower (water).[11] It also includes technologies, programs, and processes to increase energy efficiency or reduce energy consumption, such as building efficiency, more effective transmission or distribution of electricity, and vehicle technologies that improve engine efficiency or fuel economy.
- Science, which encompasses the activities of USGCRP and includes research, modeling, and observing activities to better understand climate change; efforts to analyze the cumulative effects on the environment of human activities and natural processes; and programs to provide climate information to policymakers and the public, in part to inform climate change adaptation decisions.[12] USGCRP also reports annually on funding for climate change science among certain participating agencies in its Our Changing Planet reports.[13]
- International Assistance, which uses bilateral and multilateral assistance tools to address three global climate change policy pillars: adaptation, clean energy, and sustainable landscapes.[14] For

[11] Biomass is an energy resource derived from plant- and algae-based material that includes crop wastes, forest residues, purpose-grown grasses, woody energy crops, algae, industrial wastes, sorted municipal solid waste, urban wood waste, and food waste.

[12] Climate change adaptation—defined as adjusting natural or human systems in response to actual or expected climate change—is one way for the federal government to manage climate risks.

[13] USGCRP's 13 participating agencies include the: Department of Agriculture, Department of Commerce, Department of Defense, Department of Energy, Department of Health and Human Services, Department of the Interior, Department of State, Department of Transportation, Environmental Protection Agency, National Aeronautics and Space Administration, National Science Foundation, Smithsonian Institution, and U.S. Agency for International Development. The Global Change Research Act of 1990 required the Federal Coordinating Council on Science, Engineering, and Technology to report annually to Congress the "expenditures required by each agency or department for carrying out its portion of the USGCRP," among other things. Although this reporting requirement was terminated by statute in 2000, USGCRP continues to provide this information.

[14] Sustainable landscapes sequester carbon, clean the air and water, increase energy efficiency, restore habitats, and create value through significant economic, social, and environmental benefits.

example, international assistance helps countries invest in energy efficiency programs to reduce emissions, as well as activities to help communities adapt to climate change.

OMB-Reported Climate Change Funding Increased by $4.4 Billion since 2010, but OMB Reports Do Not Provide Information about Fiscal Exposures for Budget Decisions

Annual federal climate change funding across all OMB categories increased by $4.4 billion from fiscal year 2010 through fiscal year 2017, according to OMB's three reports on such funding. However, the OMB reports did not provide information about climate-related fiscal exposures to assist policymakers in understanding budget trade-offs for climate change activities.

OMB-Reported Climate Change Funding Increased by $4.4 Billion from Fiscal Years 2010 through 2017, with Funding for Clean Energy Technology Accounting for Most of the Increase

OMB-reported annual climate change funding across all categories increased from $8.8 billion in fiscal year 2010 to $13.2 billion in proposed budget authority for fiscal year 2017 in nominal dollars (about 50 percent, or 34 percent after adjusting for inflation), with reported agency funding for clean energy technologies (i.e., technologies to reduce greenhouse gas emissions) accounting for most of the $4.4 billion increase, as shown in Figure 1.[15]

[15] Nominal dollars are not adjusted for inflation.

Table 1. OMB-Reported Increases in Climate Change Funding by Category, Fiscal Years 2010 through 2017

Dollars in millions			
	Fiscal Year		
Funding category	**2010**[a]	**2017**[b]	**Increase**
Clean energy technology	5,504	8,978	3,474
Science	2,122	2,775	653
International assistance	1,080	1,426	346
Natural resources adaptation[c]	65	NA	(65)
Adjustments for programs in multiple categories	NA	(26)	(26)
Total	**8,771**	**13,153**	**4,382**

Source: GAO analysis of Office of Management and Budget (OMB) data. I GAO-18-223.

Note: Funding is reported in nominal dollars and not adjusted for inflation.

[a]Enacted budget authority

[b]Proposed budget authority

[c]OMB has also reported on federal funding for wildlife and natural resource adaptation from fiscal years 2010 through 2013. However, the data OMB reported in the adaptation category does not fully represent adaptation funding as it only includes data from the Department of the Interior. OMB did not report funding for this category in its January 2017 report.

Specifically, reported annual funding for clean energy technology, such as nuclear, biomass, wind, and hydropower programs, increased by about $3.5 billion from fiscal years 2010 through 2017 in nominal dollars. Over the same period, OMB-reported agency funding for science and international assistance increased by $653 million and $346 million respectively, as shown in Table 1.

- Clean Energy Technology. DOE programs accounted for most of the $3.5 billion increase in annual clean energy technology funding. Specifically, funding for DOE's fusion, sequestration and hydrogen research (about $1.9 billion), energy efficiency and renewable energy programs ($656 million), as well as its carbon capture and storage and power systems research ($564 million) accounted for about 88 percent of the $3.5 billion funding increase from fiscal years 2010 through 2017, with a portion of the increase offset by decreases in other DOE programs, according to our analysis of OMB reports for this period. Funding for National

Science Foundation ($486 million) and DOD ($417 million) research accounted for another 26 percent of the funding increase.[16]

- Science. NASA programs accounted for most of the $653 million increase in annual science funding. In particular, funding for NASA's science programs accounted for about 84 percent ($546 million) of the $653 million funding increase from fiscal years 2010 through 2017, based on our analysis of OMB reports. Funding for DOE's biological and environmental research accounted for about another 12 percent ($77 million) of the funding increase.[17]
- International Assistance. Department of State and Department of the Treasury programs accounted for the greatest increases in annual international assistance funding. Specifically, State economic support funding increased by $428 million and Treasury funding for the Green Climate Fund increased by $250 million, with half of the increase offset by decreases in other programs.[18]

OMB also reported annual funding for natural resources adaptation in its August 2013 report, but did not include this category in its January 2017 report. In its August 2013 report, OMB reported funding for certain activities at Interior promoting adaptation—adjustments to natural or human systems in response to actual or expected climate change—under the natural resources adaptation category.

---

[16] Multiple agencies and programs accounted for the remaining portion of the increase. See appendixes IV and V for more information on reported funding by agency and program line item.

[17] The remaining 5 percent of the funding increase was spread across multiple agencies, such as the National Science Foundation. See appendixes IV and V for more information on reported funding by agency and program line item.

[18] According to our analysis of OMB reports, the reported funding increase for the Treasury's Green Climate Fund was offset by funding reductions across multiple programs, such as the Clean Technology Fund ($300 million), and Strategic Climate Fund ($75 million). See appendixes IV and V for more information on reported funding by agency and program line item.

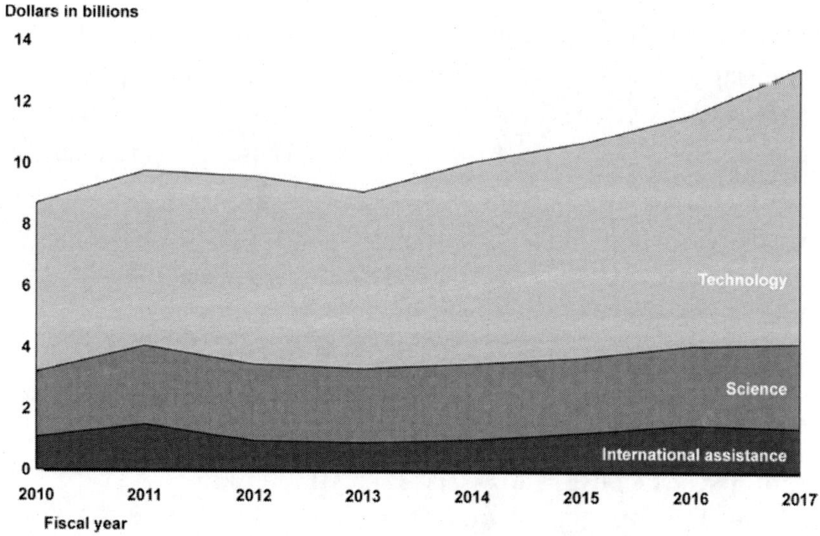

Source: GAO analysis of Office of Management and Budget (OMB) reports. | GAO-18-223.

Notes: The figure presents enacted budget authority except for fiscal years 2011, 2013, and 2017. For fiscal years 2011 and 2017, OMB reported proposed budget authority. For fiscal year 2013, OMB reported final operating level funding. Funding is reported in nominal dollars, which are not adjusted for inflation.

OMB has also reported on federal funding for wildlife and natural resource adaptation from fiscal years 2010 through 2013. However, the data OMB reported in the adaptation category does not fully represent adaptation funding as it only includes data from the Department of the Interior. OMB reported Department of the Interior funding for adaptation as follows: fiscal year 2010, $65 million; fiscal year 2011, $87 million; fiscal year 2012, $88 million. We did not include funding for this category in fiscal year 2013 because we used OMB's fiscal year 2017 report for fiscal year 2013 funding, and OMB did not report funding under this category in its 2017 report.

Funding in the American Recovery and Reinvestment Act of 2009 (Pub. L. No. 111-5 (Feb. 17, 2009)) and tax expenditures related to climate change are not included in this figure to maintain comparability with earlier GAO reports.

Figure 1. OMB-Reported Federal Climate Change Funding by Category, from Fiscal Years 2010 through 2017.

We previously reported that the Administration was developing criteria to systematically account for a broader suite of adaptation programs.[19] However, OMB staff we interviewed said that it is difficult to parse out adaptation funding for federal facilities because agencies do not necessarily separate it out from routine planning decisions. For example, an agency planning to build a sea-wall to protect a coastal facility might build it higher to accommodate rising sea levels but may not track this spending as climate change- related.

OMB Reports Are Not Clearly Linked to Climate- Related Federal Fiscal Exposure to Assist Policymakers in Understanding Budget Trade-Offs for Climate Change Activities

OMB's most recent reports to Congress on climate change funding are not clearly linked to climate-related federal fiscal exposure, although the agency has recently identified climate-related fiscal exposures in the President's budget requests and a separate assessment of the fiscal risks from climate change. Specifically, in 2015 and 2016, OMB identified areas of the federal budget as climate-related because the costs of these programs or activities were likely to increase as a direct or indirect result of climate change.[20] For example, in a section entitled the *Federal Budget Exposure to Climate Risks*, the President's budget requests for fiscal years 2016 and 2017 stated that the imprint of climate change on the federal budget was increasingly apparent in the escalating costs for several federal programs or activities. Over the last decade, the federal government incurred an estimated $357 billion in direct costs from severe weather and fire—for domestic disaster response and relief ($205 billion), crop and flood insurance ($90 billion), wildland fire management ($34 billion), and maintenance and repairs to federal facilities and federally managed lands,

[19] GAO, *Climate Change: Improvements Needed to Clarify National Priorities and Better Align Them with Federal Funding Decisions*, GAO-11-317 (Washington, D.C.: May 20, 2011).

[20] Individual agencies—such as DOD and USDA—and intergovernmental task forces have also published information about certain programs' fiscal exposures to climate risks since 2010.

infrastructure, and waterways ($28 billion).[21] While acknowledging that it was not possible to identify the portion of these costs attributable to climate change, the President's budget request stated that costs for each of these federal programs have been increasing and are inherently sensitive to climate change impacts, and can therefore be expected to continue to rise as climate change intensifies.

Additionally, in the November 2016 assessment *Climate Change: the Fiscal Risks Facing the Federal Government*, OMB and the Council of Economic Advisers estimated that recurring costs incurred by the federal government due to climate change could average about $64 billion per year across four federal program areas by late century—acknowledging key limitations and uncertainties.[22] The assessment included four federal programs or activities—wildland fire suppression, crop insurance, healthcare in response to poor air quality, and disaster relief in coastal areas—which are outside OMB's climate change funding categories of science, technology, and international assistance. In this assessment, OMB identified climate-related fiscal exposures generally aligned with three of the five areas of fiscal exposure we identified in our 2017 high risk list update: the federal government as property owner, as an insurer of property, and as a provider of disaster relief.[23]

However, the most recent OMB climate change funding reports do not include information on the funding of climate-related fiscal exposures discussed in the 2016 and 2017 Analytical Perspectives and the November

---

[21] Office of Management and Budget, *Analytical Perspectives, Budget of the United States Government for Fiscal Year 2017* (Washington, D.C.: 2016). OMB staff stated that the budget appendix provides detailed information on various appropriations and funds—including a prior year baseline for disaster-relief outlays—but does not reflect the fiscal exposures from climate change over time, such as expenditures related to climate change impacts (e.g., hazard mitigation, adaptation, and changes in disaster-relief over time attributable to climate change). The President's Budget request for fiscal year 2018 does not include a section on federal budget exposure to climate change.

[22] According to this assessment by OMB and the Council of Economic Advisers, these ranges are projections of costs that would be incurred by the federal government across four program areas given a set of assumptions on which the specific scenarios were modeled. The Executive Office of the President, *Climate Change: the Fiscal Risks Facing the Federal Government* (November 2016).

[23] GAO-17-317. The other two areas of fiscal exposure are the federal government's role as leader of a strategic plan and as the provider of technical assistance to federal, state, and other decision-makers.

2016 assessment. For example, these reports generally do not include information on the funding of actions taken by agencies to improve their program and facilities' resilience to climate change impacts. In particular, OMB stated in its August 2013 climate change funding report that federal agencies were starting to prepare for climate change through actions outlined in their Climate Change Adaptation Plans— however, federal funding associated with these actions was not included in OMB's 2017 report—based on our review.[24] Similarly, the OMB climate change funding reports we reviewed do not include funding information on other programs—such as federal domestic disaster assistance, flood insurance, and crop insurance—that OMB and others identified as climate-related fiscal exposures.

We have previously reported that a more complete understanding of fiscal exposures and the long-term effects of decisions would help policymakers make important trade-offs between spending with long-term benefits and spending with short-term benefits.[25] In May 2011, we identified broad principles or criteria for a budget process, including that it should (1) provide information about the long-term effects of decisions; (2) provide information necessary to make important trade-offs between spending with long-term benefits and spending with short-term benefits, and (3) provide for accountability and be transparent, among other principles.

Further, in October 2013 we reported that incorporating more complete information on fiscal exposures could help meet these criteria for an effective budget process. In particular, we reported that a more complete understanding of fiscal exposures and the long-term effects of decisions would help policymakers make trade-offs between spending with long-

---

[24] For example, key agencies listed under the federal government's sustainability goals to meet its energy efficiency and greenhouse gas emissions reduction targets, such as the General Services Administration, Department of Veterans Affairs and Department of Homeland Security, are not included in the OMB report. OMB instructions for implementing the March 2015 Executive Order 13693, *Planning for Federal Sustainability in the Next Decade*, directs agencies to report their progress implementing and any updates to their Climate Change Adaptation Plans as part of their annual Strategic Sustainability Performance Plan.

[25] GAO, *Budget Process: Enforcing Fiscal Choices* GAO-11-626T (Washington, D.C.: May 4, 2011) and GAO-14-28.

term benefits and spending with short-term benefits, anticipate changes in future spending, and enhance control over federal resources.

According to OMB staff, it is difficult to fully account for actions taken by agencies to improve their program and facilities' resilience, because successful preparedness efforts often involve integrating climate change considerations into existing agency programs, projects, and activities rather than establishing separate and distinct programs. Additionally, OMB staff said that they do not provide information on federal fiscal exposures to climate change in their funding report for two reasons. First, OMB staff said that OMB prefers to conduct a broader examination of federal fiscal risks as part of agencies' enterprise risk management assessments under OMB Circular A-123. Specifically, agencies are responsible for establishing practices that identify, assess, respond, and report on risks. According to the circular, these practices must be forward- looking and designed to help leaders make better decisions, alleviate threats, and identify previously unknown opportunities to improve the efficiency and effectiveness of government. OMB staff said that each agency has discretion to prioritize its risks, decide whether to consider climate change as part of this assessment, and to consider its significance as a risk to agency goals and operations. Second, OMB staff said that the agency has limited capacity and knowledge to assess climate change risks and predict future expenditures on an annual basis, due to the inherent uncertainty in climate projections. In particular, OMB staff said that its prior assessments of climate change risks were resource-intensive and reflect a previous administration's priorities.

Further, OMB staff said that they defer to Congress on whether the current funding reports are useful, and Congress has not requested information on fiscal exposures as part of OMB funding reports.

However, we and others have reported that understanding the federal government's fiscal exposure to climate change risks is increasingly critical for policymakers charged with making sound investment decisions and acting as stewards of the federal budget over the long term.[26] In

[26] For example, see GAO, *Disaster Resilience: Actions Are Underway, but Federal Fiscal Exposure Highlights the Need for Continued Attention to Longstanding Challenges*, GAO-

November 2017, we reported that DOD's overseas installations do not consistently track the costs of impacts of weather effects associated with climate change. As a result, the military services lack the information they need to adapt infrastructure at overseas installations to climate change impacts and develop accurate budget estimates for infrastructure sustainment.[27] By providing, concurrent with any future climate change funding reports to Congress, funding information for federal programs with fiscal exposure to climate change, OMB would have better assurance that it was providing policymakers with the information necessary for them to make decisions about spending trade-offs, as called for in our criteria for an effective budget process. This information should include costs to repair, replace, and improve the weather-related resilience of federally-funded property and resources; costs for federal flood and crop insurance programs; and costs for disaster assistance programs, among other identified areas of fiscal exposure to climate change.[28]

SELECTED AGENCIES' PROGRAMS WHOSE PRIMARY PURPOSE IS ADDRESSING CLIMATE CHANGE REPRESENT A SMALL PORTION OF THEIR TOTAL AGENCY-REPORTED CLIMATE CHANGE FUNDING

Based on our review of six selected agencies' budget justifications for fiscal years 2010 through 2017, we identified few federal programs whose primary purpose is to address climate change and these programs represent a small portion of their total agency-reported climate change funding.

14-603T (Washington, D.C.: May 14, 2014); National Research Council. *Disaster Resilience: A National Imperative*. (Washington, D.C.: The National Academies Press, 2012). Ekins, Paul and Speck, Stefan, 2014. "The Fiscal Implications of Climate Change and Policy Responses." *Mitigation and Adaptation Strategies for Global Change*, (Vol. 19: 355 – 374).

[27] GAO, *Climate Change Adaptation: DOD Needs to Better Incorporate Adaptation into Planning and Collaboration at Overseas Installations*, GAO-18-206 (Washington, D.C.: Nov.13, 2017).

[28] GAO-17-317 and GAO-14-28.

Table 2. Funding for Primary Purpose Programs in Selected Agencies in Our Review, Fiscal Years 2010 through 2017 Dollars in millions

Fiscal Year										
OMB category	**Primary purpose program**	**Agency**	**2010**	**2011**	**2012**	**2013**	**2014**	**2015**	**2016**	**2017**
Science	Climate Research	Commerce's NOAA	219.9	217.8	139.5	135.5	153.3	157.1	158.0	189.9
	Atmospheric Radiation Measurement Climate Research Facility	DOE	42.2	45.8	67.9	68.1	68.6	67.4	65.4	65.4
	Atmospheric System Research	DOE	26.4	27.8	26.3	26.4	26.6	26.0	26.4	26.4
	Climate and Earth System Modeling	DOE	69.1	77.9	70.7	72.9	74.0	71.1	98.7	103.5
	Terrestrial Carbon Sequestration Research	DOE	4.6	3.0	—	—	—	—	—	—
	Terrestrial Ecosystem Science	DOE	28.7	28.7	40.2	38.8	45.3	44.0	40.0	40.0
	CLARREO Pathfinder	NASA	—	—	—	—	—	—	—	19.3
	Glory Mission	NASA	31.8	12.9	—	—	—	—	—	—
	Ice, Cloud, and Land Elevation Satellite	NASA	—	3.8	0.7	—	—	—	—	—
	Ice, Cloud, and Land Elevation Satellite-2	NASA	38.9	59.7	130.5	165.9	182.2	126.5	117.4	112.4
	Orbiting Carbon Observatory-2	NASA	62.0	89.0	93.4	80.3	38.2	17.5	—	10.2
	Orbiting Carbon Observatory-3	NASA	—	—	—	7.4	16.8	1.5	—	26.3
	Solar Radiation and Climate Experiment	NASA	—	4.6	5.3	5.2	5.4	5.4	—	5.4
	Total Solar Irradiance Sensor-1	NASA	—	—	—	—	—	1.0	—	19.6
	Total Solar Irradiance Sensor-2	NASA	—	—	—	—	—	—	—	9.6
	Agroclimatology (formerly Global Change UV-B Monitoring and Research)	USDA	1.4	1.4	1.3	1.3	1.4	1.4	1.4	1.4
	Climate Change Program Office	USDA	2.9	2.2	2.2	2.0	2.5	2.7	2.6	3.0
Clean Energy Technology	Climate Change Technology Program	DOE	9.3	5.5	5.5	5.3	4.9	—	—	—
Total			**537.2**	**580.1**	**583.5**	**609.1**	**619.2**	**521.6**	**509.9**	**632.4**

Legend: DOE = Department of Energy; NASA = National Aeronautics and Space Administration; Commerce's NOAA = National Oceanic and Atmospheric Administration, Department of Commerce; USDA = U.S. Department of Agriculture, — = agencies did not report funding for the program for that year.

Source: GAO analysis of Department of Commerce, Department of Defense, Department of Energy, National Aeronautics and Space Administration, National Science Foundation, and U.S. Department of Agriculture budget justifications. | GAO-18-223.

Notes: Funding is reported in nominal dollars and not adjusted for inflation. For two of the six agencies—DOD and NSF—we did not identify any primary purpose climate change programs based on program descriptions in their budget justifications.

In particular, we identified 18 out of 533 programs in agency budget justifications that met our definition for a primary purpose climate change program, based on the program descriptions in the budget justifications.[29] Seventeen of the 18 primary purpose programs we identified fell under OMB's science category, and one primary purpose program fell under OMB's clean energy technology category (see Table 2), according to our analysis of agency budget justifications and OMB reports.

Under the science category, 13 of the 18 programs we identified focused on collecting observations of climate-related variables, such as atmospheric carbon dioxide concentrations and polar ice thickness, to enhance scientific understanding of climate change. The five remaining primary purpose programs we identified focus on other aspects of climate change, such as modeling and providing climate information to stakeholders to facilitate adaptation. According to USGCRP officials, these programs conduct research to support individual agency missions, such as agriculture, and address climate change as a cross-cutting issue, rather than a primary purpose. Appendix I contains further information on the 18 primary purpose programs we identified.

The primary purpose programs we identified represent a small portion of total reported climate change funding for these agencies. For the 18 primary purpose programs we identified, agencies reported about $632 million in proposed funding for fiscal year 2017 in their budget justifications, which represents about 6 percent of the $11.4 billion in total proposed funding to address climate change that OMB reported for these agencies in fiscal year 2017.[30] The majority of the selected agencies' reported funding supports programs where climate change is one of multiple purposes being addressed.

---

[29] We reviewed the budget justifications of six agencies—Commerce, DOD, DOE, NASA, NSF, and USDA—that accounted for $11.4 billion of the $13.2 billion in fiscal year 2017 funding to address climate change as reported by OMB. For two of the six agencies— DOD and NSF—we did not identify any primary purpose climate change programs based on program descriptions in their budget justifications.

[30] Due to differences in the timing of agencies' budget justification submissions and OMB's climate change funding reports, agencies and OMB reported different types of funding for each fiscal year other than 2017. For example, for fiscal year 2013, OMB reported funding as "final operating level" whereas agencies generally reported funding for fiscal year 2013 as "actual budget authority."

Based on our analysis of agency budget justifications and other documents, most programs (515 of the 533) we reviewed did not meet our definition of a primary purpose climate change program because they were broad-based technology and science programs where climate change is one of multiple purposes.

For example, 331 of the 515 multiple-purpose programs we identified focus on advancing clean energy technologies, primarily for purposes other than addressing climate change, such as: reducing energy costs, increasing energy security by shifting away from foreign fuel sources, advancing economic development, and improving air quality. Many of these programs are located in DOE, and the 1977 law establishing the agency included congressional findings that the country's increasing shortage of nonrenewable energy resources and increasing dependence on foreign energy supplies presented a serious threat to national security, among other things.[31] One purpose of the law that established DOE was to place a major emphasis on the development and commercial use of solar, geothermal, recycling, and other technologies utilizing renewable energy resources. Clean energy technology programs at the six selected agencies accounted for roughly $8.7 billion of $13.2 billion (about 66 percent) in proposed funding for fiscal year 2017, as reported by OMB.

Additionally, 137 of the 515 multiple-purpose programs conduct or support climate-related scientific research that agency budget justifications and other documents described as serving multiple purposes, such as weather forecasting or understanding coastal ecosystems in the context of multiple factors. For example, several National Oceanic and Atmospheric Administration (NOAA) programs contribute to observing and modeling the earth's climate, as part of NOAA's weather and severe storm forecasting responsibilities, to ensure the safety of U.S. citizens, public property, and infrastructure.[32] Additionally, the NSF contributed to the USCGRP research program through its broad-based Science, Engineering, and Education for Sustainability Program, which provides funding for

---

[31] 42 U.S.C. § 7111.

[32] For example, NOAA collects observational data on a variety of characteristics, such as weather patterns, precipitation, ocean temperature, and coastal ecosystem characteristics, using satellite and earth-based instruments.

research projects on the sustainability of coastal systems in the context of multiple stressors—such as increasing development—in addition to climate change-related stressors such as sea-level rise and ocean acidification.

The remaining 47 multiple-purpose programs we identified focus on activities such as adaptation or international assistance (23 programs) or activities that fell into more than one OMB category (24 programs). For example, we identified a USDA program for integrated resource restoration that promotes watershed resilience through forest management, range management, wildlife and fisheries habitat management, and other activities. We also identified a DOD program on sustainable ranges and lands that provides ecosystem vulnerability assessment and ecosystem analysis, observing, modeling and mitigation technologies to support sustainable, realistic access and use of the Army's ranges and lands. Neither the integrated resource restoration program nor the sustainable ranges and lands program falls within one of OMB's climate change funding categories.

Selected Agencies' Programs Whose Primary Purpose Is Addressing Climate Change Are Fragmented but Serve Different Purposes, Audiences, Times, and Scales

The 18 primary purpose climate change programs we identified from selected agency budget justifications are fragmented across multiple federal agencies that address climate change in ways specific to their agency missions.[33] However, we also found that the 18 primary purpose programs we identified serve different purposes, target different audiences, or operate at different times and scales, among other reasons, which minimize potential overlap or duplication among these programs.

[33] See appendix I for additional information about the 18 primary purpose programs we identified.

Additionally, USGCRP officials and program managers collaborate to avoid potential negative effects of fragmentation, according to USGCRP officials we interviewed and National Academies' reports.

Primary Purpose Climate Change Programs Are Fragmented Across the Selected Federal Agencies and Reflect Three Types of Climate Change Activities

The 18 primary purpose programs we identified are fragmented across four of the six agencies we reviewed, and address climate change in the context of their agency missions. These programs conduct activities that generally fall into three broad climate change areas—observing, modeling, and facilitating adaptation to climate impacts (see Table 3).[34] See appendix I for additional information about the 18 primary purpose programs.

Table 3. Focus Area of Six Selected Agencies' 18 Programs with the Primary Purpose of Addressing Climate Change

Program Focus						
OMB category	**Primary purpose program**	**Agency**	**Climate observing**	**Climate modeling**	**Climate adaptation activities**	**Other**
Science	Atmospheric Radiation Measurement Climate Research Facility	DOE	X			
	Terrestrial Ecosystem Science	DOE	X			
	Atmospheric System Research	DOE		X		
	Climate and Earth System Modeling	DOE		X		
	Terrestrial Carbon Sequestration Research[a]	DOE				X
	Climate Research	Commerce's NOAA	X			
	CLARREO Pathfinder	NASA	X			
	Glory Mission[a]	NASA	X			
	Ice, Cloud, and Land Elevation Satellite[a]	NASA	X			

[34] As mentioned previously, fragmentation refers to those circumstances in which more than one federal agency (or more than one organization within an agency) is involved in the same broad area of national need and opportunities exist to improve service delivery (GAO-15-49SP).

Table 3. (Continued)

Program Focus						
OMB category	**Primary purpose program**	**Agency**	**Climate observing**	**Climate modeling**	**Climate adaptation activities**	**Other**
	Ice, Cloud, and Land Elevation Satellite-2[b]	NASA	X			
	Orbiting Carbon Observatory-2[b]	NASA	X			
	Orbiting Carbon Observatory-3[b]	NASA	X			
	Solar Radiation and Climate Experiment	NASA	X			
	Total Solar Irradiance Sensor-1[b]	NASA	X			
	Total Solar Irradiance Sensor-2[b]	NASA	X			
	Agroclimatology (formerly Global Change UV-B Monitoring and Research)	USDA	X			
	Climate Change Program Office	USDA			X	
Clean energy technology	Climate Change Technology Program[a]	DOE				X
Total		**18**	**13**	**2**	**1**	**2**

Legend: DOE = Department of Energy; NASA = National Aeronautics and Space Administration; Commerce's NOAA = National Oceanic and Atmospheric Administration, Department of Commerce; USDA = U.S. Department of Agriculture.

Source: GAO analysis of Department of Commerce, Department of Defense (DOD), Department of Energy (DOE), National Aeronautics and Space Administration (NASA), National Science Foundation (NSF), and U.S. Department of Agriculture (USDA) budget justifications and program documents. | GAO-18-223.

Notes: GAO identified programs in the budget justifications of four of six selected federal agencies that the agencies described as having climate change as their primary purpose. Several of these programs focus on a particular program area, but conduct research that contributes to multiple program areas, such as DOE's Atmospheric Radiation Measurement Climate Research Facility, which primarily conducts climate observing, but also includes a modeling component. Additionally, several primary purpose programs award research grants.

[a]These programs are no longer listed as separate line items in agency budget justifications.

[b]These NASA programs are follow-on projects that will replace satellites that are approaching or have reached the end of their mission lifespan.

Observing

Thirteen primary purpose programs we identified within NASA, DOE, NOAA, and USDA collect a variety of climate data in support of their respective agency missions. Specifically, nine NASA earth science missions—two satellites of which were operating in fiscal year 2017—have collected or plan to collect climate observing data, according to our review of NASA budget justifications and program documents.[35] These data include changes in polar ice thickness, varying atmospheric carbon dioxide concentrations, and the effects of aerosols on climate change, as measured from space using satellites and instruments on the International Space Station. NASA provides these data to a wide community of users, including other federal agencies, international partners, academia, and the public, to better understand the Earth system and its response to natural and human-induced changes, according to NASA's 2017 budget justification. Additionally, two DOE programs collect climate data, such as observations on the cycling of carbon between terrestrial ecosystems and the atmosphere, as well as long-term observations of atmospheric variables, such as aerosols, clouds, and incoming solar radiation to improve climate projections and inform the development of solutions to U.S. energy and environmental challenges.[36] One NOAA program collects atmospheric and oceanic observations to describe and understand the state of the oceans and climate variability across multiple time-scales. Lastly, one USDA program monitors ultraviolet radiation (UV-B) reaching the Earth's surface over wide geographic areas of the United States, to

---

[35] The two operating satellites are the Orbiting Carbon Observatory-2, which launched in 2014, and the Solar Radiation and Climate Experiment satellite, which launched in 2003. The Total Solar Irradiance Sensor-1 launched in December 2017 and is expected to start returning data in February 2018. Four satellites—the CLARREO Pathfinder, the Ice, Cloud, and Land Elevation Satellite-2, the Orbiting Carbon Observatory-3, and the Total Solar Irradiance Sensor- 2—are under development as follow-on missions to existing satellites or planned satellites and slated to fly in 2018 or later. Two satellite missions—the Glory Mission and the Ice, Cloud, and Land Elevation Satellite were no longer listed as separate line items in NASA's budget justification as of 2011 and 2010, respectively.

[36] According to USGCRP officials, its budget cross-cut includes the basic research activities of these DOE programs, not the application of their research by other DOE programs or external scientists.

improve understanding of ultraviolet radiation impacts and climate change on agriculture.

Modeling

Adaptation

Two primary purpose programs we identified within DOE focus on climate modeling, although other primary purpose programs we reviewed also include a modeling component in their research. DOE's Atmospheric System Research program uses the agency's climate observing data to model atmospheric processes and develop model parameters to help reduce uncertainty in climate change models and projections. DOE's Climate and Earth System Modeling program develops model components as well as earth system models that include human impacts and natural systems to simulate climate variability and predict change at regional and global scales. Improved climate projections from these programs help inform DOE's energy supply and infrastructure decisions, as well as the development of solutions to environmental challenges, according to DOE program documents.[37] Additionally, NOAA's Climate Research program includes a modeling component that uses mathematical models and high-performance computer simulations to understand atmosphere, ocean, biosphere, and cryosphere dynamics and to make projections about future marine ecosystems, atmospheric composition, and air quality across multiple time scales. DOE's Atmospheric Radiation Measurement (ARM) Climate Research Facility has a modeling component that conducts high resolution modeling of atmospheric and cloud processes using the facility's observational data.

DOE, NOAA, and other scientists use the model outputs to better understand cloud formation, earth system science and severe weather.[38]

---

[37] Ibid.

[38] Other primary purpose programs also use modeling or develop modeling components. For example, DOE's Terrestrial Ecosystem Science program uses Earth system models to identify research gaps and design field experiments to address the identified uncertainties. USDA's Agroclimatology program develops modeling tools that integrate observing data, ultraviolet effects studies, and satellite observations to study how climate and crop

Two primary purpose programs we identified provide climate information to stakeholders, in part to facilitate adaptation efforts. USDA's Climate Change Program Office (CCPO) provides data, tools, and pertinent information to stakeholders on relevant climate change impacts, such as water availability and wildfires. Additionally, NOAA's Climate Research program has a regional climate data and information component that incorporates research into information and products, ranging from short- term weather forecasts to longer-term climate forecasts and assessments to improve the ability of decision makers and communities to plan for and respond to climate variability and change. Both of these programs also coordinate climate research internally and externally, among other activities.

Other Activities

Two DOE primary purpose programs we identified focused on reducing greenhouse gas concentrations in the atmosphere. For example, DOE's Terrestrial Carbon Sequestration Research program conducted research on how to enhance the carbon storage potential of soil and vegetation through economically competitive methods, such as applying coal- combustion byproducts to soil to absorb additional carbon dioxide.

Additionally, DOE's Climate Change Technology program supported efforts to accelerate the development and reduce the cost of new and advanced technologies that could avoid, reduce, or capture and store greenhouse gas emissions and increase economic growth. Both DOE programs are no longer listed as separate line items in DOE's budget justifications.[39]

---

production interact and associated effects on management practices and agricultural economics.

[39] DOE's Terrestrial Carbon Sequestration Research program was not listed as a separate line item in DOE's budget justification after fiscal year 2011 and DOE's Climate Change Technology Program was not listed as a separate line item in DOE's budget justification after fiscal year 2014.

Programs Serve Different Purposes, Audiences, Times, and Scales, and Interagency Collaboration Helps Avoid Potential Negative Effects from Fragmentation

We found that the 18 primary purpose programs we identified serve different purposes, target different audiences, or operate at different time periods and global, regional, or local scales. Further, we found that USGCRP and agency program managers collaborate to help avoid potential negative effects from fragmented climate change programs.

Therefore, although fragmented, the 18 programs we reviewed do not show signs of overlap.

Different Purposes, Audiences, Times, and Scales

We found that the 18 primary purpose programs we identified serve different purposes, target different audiences, or operate at different times and scales, among other reasons, which minimizes potential overlap or duplication among these programs. For example, primary purpose programs we identified across DOE, NASA, NOAA and USDA all collect climate observing data related to solar radiance in ways specific to their missions. In particular:

The ARM Climate Research Facility (a DOE program) collects observational data on aerosols, radiation, clouds, and atmospheric conditions across different climatic regimes to help resolve uncertainties in climate and earth system models. According to agency documents, DOE uses these models for energy and related infrastructure planning. For example, DOE models water availability for energy applications, temperature extremes affecting energy use, and predicting storm and other weather-related impacts on energy delivery.[40]

---

[40] According to USGCRP officials, its budget cross-cut includes the basic research activities of these DOE programs, not the application of their research by other DOE programs or external scientists.

- NASA satellites, such as the Total Solar Irradiance Sensor,[41] take absolute measurements of the sun's energy input to earth, and the distribution of the sun's energy input across ultraviolet, visible, and other wavelengths of light to quantify solar variations and their effect on the atmosphere and climate. For example, these data enable scientists to study the sun's natural influence on Earth's ozone layer in support of NASA's programmatic mission to monitor and assess the health of the Earth's stratosphere. Scientists also use the data to understand the sun's influence on atmospheric circulation, clouds, and ecosystems.
- NOAA's Climate Research program monitors and models atmospheric and oceanic conditions to understand the state of the oceans, predict climate variability, and develop a variety of products. For example, NOAA uses these data to develop short-term weather forecasts, early warning systems, and longer-term climate forecasts and assessments to improve the ability of decision makers and communities to plan for and respond to climate variability and change, such as changes in sea-level and extreme weather events. [42]
- USDA's Agroclimatology Program—administered by USDA's National Institute of Food and Agriculture (NIFA)—monitors surface-level ultraviolet and photosynthetically-active radiation and studies its effects on different crop varieties, in accordance with NIFA's mission to invest in and advance agricultural research, education, and extension. Scientists use these data to understand the interactions between crops, climate, and ultraviolet radiation, and assess the effects on management practices and agricultural economics.

[41] According to program documents, NASA's Total Solar Irradiance Sensor missions measure the sun's energy input to Earth. Various satellites have captured a continuous record of this solar energy input since 1978.

[42] For example, NOAA climate data and early warning systems help communities reduce vulnerability to extreme weather; prepare for drought and water resource challenges; manage risks to coastlines and coastal infrastructure; sustainably manage marine ecosystems; and adapt to and mitigate climate impacts, according to agency documents.

Some of the primary purpose programs we identified also target different audiences. In particular, both NOAA's Climate Research program and USDA's Climate Change Program Office (CCPO) provide information to help inform decisions about adapting to climate change, but generally target different audiences. For example:

- NOAA's Climate Research program has a regional climate information services component that provides information and products to improve decision makers' ability to prepare for and respond to short and long-term climate variability and change.[43] The program tailors its information on climate impacts based on regional stakeholder input to help stakeholders and communities expand their capacity to prepare for and respond to floods, drought, wildfire, extreme heat, changes in water supply and snowpack, melting permafrost, sea level rise, and storms.
- USDA CCPO's Regional Climate Hubs program has incorporated data, tools and forecasts from USDA agencies and partners—including NOAA—into integrated services specific to the agricultural and forestry sectors, such as climate impacts on crop varieties, forest health and management, grazing lands, and rural communities.[44] USDA tailors this information to support climate-informed decision- making by farmers, ranchers, and forest landowners, with the goal of maintaining agricultural production, managing natural resources, and supporting rural economic development under increasing climate variability.

Further, some primary purpose programs provide data that supplement research by other programs or that are necessary to ensure data validity. For example, DOE's Atmospheric System Research (ASR) program conducts short-term, targeted field campaigns on specific atmospheric

[43] Examples of decision makers include water utilities, USDA agricultural extension agents, regional, state, and local planners and communities, among others.
[44] For example, NOAA and USDA collaborate on relevant tools, such as the National Integrated Drought Information System.

processes to supplement the long-term observational data collected by the ARM Climate Research Facility, which ASR scientists use to develop model parameters for clouds and aerosols. In turn, scientists within DOE's Climate and Earth System Modeling program incorporate ASR's model parameters and observational data the ARM Climate Research Facility and Terrestrial Ecosystem Science program into its modeling to reduce uncertainty in model representations of atmospheric chemistry, aerosols, and the interactions between ecosystems and climate.

Additionally, according to the 2014 *National Plan for Civil Earth Observations*, USGCRP, and the National Academies, airborne, terrestrial, and marine observations supplement satellite observations because they provide high degrees of resolution and density and are essential to validate satellite-derived data products.[45]

Moreover, we determined that primary purpose programs with similar missions were not overlapping or duplicative because they conducted activities over different time periods, at different scale, or collected different measurements. Specifically:

- Within NASA, several primary purpose programs focused on conducting similar observing activities across different time periods. For example, five of NASA's nine primary purpose programs—Ice, Cloud, and Land Elevation Satellite 2, Orbiting Carbon Observatory 2 and 3, and Total Solar Irradiance Sensors 1 and 2—are follow-on missions that are to replace previous or current satellites to maintain the continuity of observational data when the current satellites no longer function. Due to the long lead-time necessary to develop satellites, the development stage of

[45] Office of Science and Technology Policy, *National Plan for Civil Earth Observations.* (Washington, D.C.: July 18, 2014); U.S. Global Change Research Program, *National Global Change Research Plan 2012 – 2021: A Strategic Plan for the U.S. Global Change Research Program* (USGCRP) (Washington, D.C.: 2012); U.S. Global Change Research Program, *National Global Change Research Plan 2012 – 2021: The Triennial Update* (Washington, D.C.: 2017); and The National Academies of Sciences (NAS), Engineering, and Medicine, *Accomplishments of the U.S. Global Change Research Program* (Washington, D.C.: 2017).

follow-on missions may be concurrent with the operational stage of existing satellites.

- NASA's Orbiting Carbon Observatory satellites make space-based measurements of atmospheric carbon dioxide concentrations at regional and global scales to increase our understanding of how carbon dioxide sources and sinks are geographically distributed, as well as how carbon sink efficiencies are changing over time.[46] In contrast, DOE's Terrestrial Ecosystem Science program manages the AmeriFlux network, an interagency effort which measures surface flows of carbon primarily at sites in North America, and conducts targeted field studies on the cycling of carbon, nutrients, and water within specific ecosystems that are poorly understood in models— such as arctic and tropical ecosystems—according to agency documents.[47] Further, although DOE's Terrestrial Carbon Sequestration Research program also conducted research to understand carbon sequestration, it researched how land-based ecosystems' carbon sequestration productivity could be enhanced by changes in land-use patterns, or the addition of absorptive materials to soil, such as coal by-products, among others.

Collaboration

USGCRP officials identified various collaboration methods that USGCRP officials and agency program managers use—such as interagency strategic planning, jointly-funded projects, and routine meetings of relevant officials—that help to minimize potential overlap and duplication among climate change programs.[48] For example, USGCRP's

---

[46] A carbon sink reabsorbs carbon dioxide from the atmosphere and stores it on Earth. For example, trees remove carbon dioxide from the atmosphere through photosynthesis and oceans absorb carbon dioxide.

[47] DOE, NASA, NOAA, and USDA's Forest Service supported the formation of the AmeriFlux network, which measures carbon flows at over 120 sites.

[48] In September 2012, GAO identified several key considerations for implementing interagency collaboration mechanisms. For the purposes of this chapter, we did not fully assess USGCRP's efforts against these considerations; however the interagency collaboration methods that we identified are consistent with our prior work. For more information, see GAO, *Managing for Results: Key Considerations for Implementing Collaboration Mechanisms*, GAO-12-1022 (Washington, D.C.: Sept. 27, 2012).

strategic planning process includes working with agency program managers to establish research priorities across the 13 participating agencies.[49] Additionally, NASA and NOAA are statutorily required to submit a joint annual report to relevant congressional committees describing how their earth science programs, which include their climate-related earth observation efforts, will be coordinated.[50]

Some agencies also jointly-fund projects and collaborate on areas of mutual interest. For example, DOE's Atmospheric Radiation Measurement (ARM) Climate Research Facility, NASA and NOAA collaborated on a field campaign to study how aerosols and atmospheric rivers—regions in the atmosphere that are responsible for most of the horizontal transport of water vapor outside of the tropics—contribute to forming snowpacks that supply most of the water in the western United States.[51] According to NOAA, the data will help improve short and long term predictions of precipitation, and help decision makers prepare for extreme precipitation, hazard response, and management of water supply, among other uses. Additionally, we found that although NOAA, NASA, DOE, and USDA's climate observing efforts are designed for different purposes, have different geographic coverage, and take measurements using different instruments, these agencies also collaborate on climate observing activities in areas of mutual interest, such as the Arctic.[52]

Further, USGCRP officials we interviewed said that agency program managers meet regularly to coordinate their efforts for addressing climate change, including through USGCRP interagency working groups focused

---

[49] U.S. Global Change Research Program, *National Global Change Research Plan 2012 – 2021.* (2012).

[50] 51 U.S.C. § 60505(b).

[51] Other partners include the Scripps Institution of Oceanography and Naval Research Laboratory.

[52] For example, DOE's ARM Facility and NOAA's Global Monitoring Division network collaborate on measurements at NOAA's Barrow, Alaska site, and DOE's Southern Great Plains site. NOAA's Climate Reference Network and USDA's Agroclimatology network both use observations collected by USDA's Agricultural Research Service at its sites in Las Cruces, N. Mex. and Nunn, Colo. NOAA also collaborated with USDA and the University of Alaska-Fairbanks for its sites in Alaska, and both USDA and NOAA collaborated with the Department of the Interior's National Park Service for their observing sites in Big Bend National Park.

on specific issues, such as integrative modeling.[53] For example, the officials said that the program managers use meetings to facilitate awareness about different climate change models. Additionally, in response to a 2012 report by the National Research Council that recommended greater coordination and more consistent evaluation of U.S. climate models, USCGRP convenes annual U.S. climate modeling summits to facilitate coordination on specific items of shared interest, such as joint modeling activities.[54]

According to a recent report on USGCRP accomplishments by the National Academies, coordination efforts facilitated by USGCRP reduced potential duplication, improved efficiency, and enabled key advances.[55] Specifically, the National Academies found that USGCRP's coordination efforts enabled key advances in developing global earth observation systems, improving modeling, and incorporating advances in understanding relevant processes to carbon cycle research, among others. For example, USGCRP's Carbon Cycle Interagency Working Group helps coordinate the intensive, interagency field campaigns necessary to unite ground-based, airborne, and satellite-based observations to understand carbon cycle variability and change. The working group supports an interagency call for complementary research proposals across USDA, DOE, NASA, and NOAA to improve the understanding of changes in the distribution and cycling of carbon between the land, ocean, and atmosphere.[56] Nonetheless, the National Academies and others have

---

[53] Integrated models are advanced models that include the physical, chemical, biological, and human components of the Earth system, as well as the feedbacks among them, to provide more comprehensive and realistic modeling of global change. Other USGCRP interagency working groups include process research, observations, carbon cycle, climate change and human health, education, international activities, sustained assessment, indicators, scenarios, global change information, social-sciences coordination, and adaptation science, according to USGCRP's 2017 update to its strategic plan.

[54] Moss et al., 2016. *Understanding Dynamics and Resilience in Complex Interdependent Systems: Prospects for a Multi-Model Framework and Community of Practice*. Report of a workshop held under the auspices of the USGCRP Interagency Group on Integrative Modeling, with support from DOE. (December, 2016.)

[55] The National Academies of Sciences, Engineering, and Medicine, *Accomplishments of the U.S. Global Change Research Program.* (2017).

[56] Another coordination effort, under the Interagency Arctic Research Policy Committee, leveraged NASA and DOE observations and experiments to provide complementary ground and airborne measurements of ecosystem responses to changing conditions.

emphasized the need for continued collaboration among federal agencies regarding their various climate change science programs and activities, such as climate modeling.[57]

However, as we noted earlier, the three funding categories of science, technology, and international assistance in recent OMB funding reports generally do not capture all climate change-related programs and activities, such as actions taken by agencies to increase their resilience to climate change impacts and programs with climate-related fiscal exposure, such as disaster assistance. Additionally, we found that OMB reports and most agency budget justifications we reviewed do not provide sufficient detail to identify and track all climate-related programs and activities over time. In particular, most agency budget justifications we reviewed provided detailed information on climate change activities under multi-purpose programs, but generally did not report funding for these climate change activities as separate line items in their budget justifications or as line items in OMB's funding reports.[58]

Further, we reported in October 2014 that there is no comprehensive list of federal programs and funding information—including for government- wide efforts such as climate change—which would help policymakers determine the scope of the federal government's involvement in particular areas, and where action is needed to address crosscutting issues, including instances of fragmentation, overlap, or duplication.[59] We made several recommendations to OMB to ensure the usefulness of a federal program inventory, and provide greater transparency and

---

[57] National Research Council of the National Academies, *A National Strategy for Advancing Climate Modeling* (Washington, D.C.: 2012).

[58] For example, the January 2017 OMB report includes funding for USDA's National Institute for Food and Agriculture (NIFA) under its science category, but does not include further detail on which NIFA programs were included in the reported funding. When reviewing NIFA's budget justification, we found several climate change activities under the Agriculture and Food Research Initiative and other programs, but NIFA did not track actual costs for these activities as line items over time.

[59] GAO, *Government Efficiency and Effectiveness: Inconsistent Definitions and Information Limit the Usefulness of Federal Program Inventories*, GAO-15-83 (Washington, D.C.: Oct. 31, 2014). In a September 2017 report, we identified a series of iterative steps that can be used to develop an inventory and potential benefits. GAO, *Federal Programs: Information Architecture Offers a Potential Approach for Development of an Inventory*, GAO-17-739 (Washington, D.C.: Sept. 28, 2017).

consistency in federal program funding and performance information; however, as of January 2018, OMB had not implemented them.[60]

According to OMB staff, its climate change funding reports provide a sufficient level of detail for Congress to identify fragmentation, overlap, and duplication within climate change funding. OMB also assesses potential fragmentation, overlap, and duplication government-wide as part of its overall budget process, contained most recently in its Major Savings and Reform volume of the President's budget request for fiscal year 2018.[61] Additionally, OMB staff said that the U.S. already prepares an inventory of federal climate change programs as part of its biennial Climate Action Report to the U.N. Framework Convention on Climate Change.[62] Further, OMB staff said that they are taking GAO's previous recommendations—to develop a federal program inventory to provide greater transparency and consistently in federal program funding and performance information—into account as part of the broader implementation of the Digital Accountability and Transparency Act of 2014 and GPRA Modernization Act of 2010.[63]

However, as we previously mentioned, several agency budget justifications we reviewed provided detailed information on climate change activities under multi-purpose programs, that were not identifiable based upon the level of detail within OMB's funding reports. Additionally, GAO's specific criteria for identifying fragmented, overlapping, and duplicative programs includes a detailed comparison of programs across several characteristics and a comprehensive assessment of potential

---

[60] For example, we recommended that OMB revise its guidance to direct agencies to collaborate with each other in defining and identifying programs that contribute to common outcomes, provide complete performance information, and to consult with external stakeholders on their program inventories, among other recommendations.

[61] Office of Management and Budget, *Major Savings and Reform Volume, Budget of the United States Government for Fiscal Year 2018* (Washington: 2017). Under the previous administration, OMB's assessment of overlapping and duplicative funding was reported as part of the President's Budget in a section titled, *Cuts, Consolidations, and Savings*.

[62] The U.S. Department of State published the most recent Climate Action Report in 2016, which can be accessed at: http://unfccc.int/national_reports/biennial_reports_and_iar/submitted_biennial_reports/item s/7550.php.

[63] Digital Accountability and Transparency Act of 2014, Pub. L. No. 113-101, 128 Stat. 1146 (May 9, 2014); GPRA Modernization Act of 2010, Pub. L. No. 111-352, 124 Stat.3866 (Jan. 4, 2011).

positive and negative effects of fragmentation, overlap, and duplication, but these detailed comparisons are generally not included in OMB's justifications for eliminating programs in its fiscal year 2018 Major Savings and Reform volume.[64] Moreover, the 2016 Climate Action Report prepared for the U.N. Framework Convention on Climate Change provides information on U.S. progress toward meeting its emission reduction targets and international assistance activities, rather than a comprehensive examination of federal climate change programs, according to our analysis of the 2016 report. Further, because our review focused on six selected agencies' programs based on OMB-reported funding rather than a comprehensive list of federal climate change programs, and OMB reports do not include information on programs with climate-related fiscal exposures, there may be additional opportunities to identify potential fragmentation, overlap, or duplication for federal climate change efforts.

Since 2011, we have reported that identifying and addressing fragmentation, overlap, and duplication have yielded several benefits, such as improved government efficiency, improved program effectiveness, and increased assurance that programs comply with laws and funds are legally spent.[65] We have also reported that increasing interagency collaboration on crosscutting issues may help reduce fragmentation, overlap, and duplication among programs that have been added incrementally over time to address new challenges.[66] Because OMB collects and reports information on federal climate change funding, OMB is uniquely situated to conduct an assessment of potential fragmentation, overlap, or duplication across the full range of agencies engaged in climate change

---

[64] OMB identified 36 programs for elimination or reduction based on fragmentation, overlap, or duplication in the *Major Savings and Reforms* volume of the President's budget request for fiscal year 2018. For 6 of the 36 programs, OMB referenced prior GAO analyses of fragmentation, overlap, and duplication in its justification. We could not determine what criteria were applied to identify fragmentation, overlap, or duplication for the remaining 30 programs because OMB did not include a detailed analysis. For more information on GAO's criteria for evaluating fragmentation, overlap, and duplication among programs, see GAO-15-49SP.

[65] For example, our most recent report on government-wide fragmentation, overlap, and duplication is GAO-17-491SP.

[66] GAO, *Managing for Results: GPRA Modernization Act Implementation Provides Important Opportunities to Address Government Challenges*. GAO-11-617T (Washington, D.C.: May 10, 2011) and GAO-12-1022.

activities. By providing a detailed analysis of such areas to Congress, in conjunction with OMB's funding reports, OMB could help decision makers more effectively target limited resources.

CONCLUSION

Since 1993, the federal government has reported over $154 billion in funding for activities to understand and address climate change, and annual funding has increased over time, according to OMB climate change reports to Congress. Given the fiscal constraints facing the federal government, we have previously reported that a more complete understanding of fiscal exposures and the long-term effects of decisions would help policymakers make important trade-offs between spending with short-term and long-term benefits. However, the most recent OMB climate change funding reports do not include funding information on federal programs with significant fiscal exposures to climate change identified by OMB and others—such as federal disaster assistance, flood insurance, and crop insurance. By providing, concurrent with any future climate change funding reports to Congress, funding information for federal programs with fiscal exposure to climate change, OMB would have better assurance that it was providing policymakers with the information necessary for them to make decisions about spending trade- offs. This information should include costs to repair, replace, and improve the weather-related resilience of federally-funded property and resources; costs for federal flood and crop insurance programs; and costs for disaster assistance programs, among other identified areas of fiscal exposure to climate change.

Additionally, based on our review of six selected agency budget justifications, funding for programs whose primary purpose is to address climate change represents a small portion of total agency-reported climate change funding to OMB, and is fragmented across multiple agencies, although we did not find signs of overlap. However, there may be additional opportunities to identify potential fragmentation, overlap, or duplication for federal climate change efforts because several agency budget justifications we reviewed provided information on climate change activities under multi-purpose programs that were not identifiable within OMB's funding reports, and detailed analyses of overlapping and duplicative programs are generally not included in OMB's justifications for eliminating programs, among other reasons. We have previously reported that identifying and addressing fragmentation, overlap, and duplication have yielded several benefits, such as improved program effectiveness, increased collaboration, and improved government efficiency by eliminating or avoiding unnecessary overlap and duplication. By providing, concurrent with any future climate change funding reports to Congress, a detailed analysis of overlapping or duplicative federal climate change programs to Congress, OMB could help decision makers more effectively target limited resources.

RECOMMENDATIONS FOR EXECUTIVE ACTION

We are making the following two recommendations to OMB:

The Director of OMB should provide, concurrent with any future climate change funding reports to Congress, funding information for federal programs with fiscal exposure to climate change. This information should include costs to repair, replace, and improve the weather-related resilience of federally-funded property and resources; costs for federal flood and crop insurance programs; and costs for disaster assistance programs. (Recommendation 1)

The Director of OMB should provide, concurrent with any future climate change funding reports to Congress, a detailed analysis of federal climate change programs it considers to be fragmented, overlapping, or duplicative. (Recommendation 2)

Agency Comments and Our Evaluation

We requested comments on a draft of this product from CEQ, OMB, the Office of Science and Technology Policy, and USGCRP within the Executive Office of the President. CEQ and the Office of Science and Technology Policy did not provide comments. USGCRP provided technical comments, which we incorporated into the draft as appropriate. OMB staff provided oral comments on the draft report. In their comments, OMB staff said that they generally agreed with our findings. However, they disagreed with our recommendations because the staff said that they believe existing budget processes are sufficient for identifying significant fiscal risks to agency operations, including climate change, and for identifying climate change programs that are fragmented, overlapping, or duplicative.

OMB staff said that they disagreed with our recommendation for OMB to provide Congress with funding information on federal programs with fiscal exposure to climate change, concurrent with future OMB funding reports because OMB prefers to conduct a broader examination of federal fiscal risks during the budget process. Specifically, OMB staff said that OMB Circular A-123 directs agencies to conduct enterprise risk management assessments to identify significant risks to agency goals and operations and that agencies have discretion on how to prioritize identified risks and whether climate change should be included. Additionally, outside of the existing budget process, OMB staff said that there is significant uncertainty in climate projections, that the ability to distinguish impacts and decisions associated with climate change from other drivers is often not possible, and that any analyses to evaluate those distinctions or assess future trends in expenditures would require substantial resources. Further, OMB staff said that no test exists to identify climate change programs with fiscal exposures and that its identification of such programs would be subjective and not transparent.

OMB staff suggested that we revise our first recommendation to reflect that agencies should consider fiscal exposure to climate change as part of their enterprise risk management assessments under OMB Circular A-123.

We agree with OMB that it is important to conduct a broad examination of federal fiscal risks during the budget process and that agencies should consider fiscal exposure to climate change as part of their enterprise risk management assessments under OMB Circular A-123. However, we were asked to review reported federal funding for climate change activities, and evaluating agencies' enterprise risk management assessments was outside the scope of our review. For this chapter, we examined the extent to which OMB's climate change funding reports were clearly linked to federal fiscal exposure to climate change and we found that they were not. We also found that OMB would have better assurance that it was providing policymakers with the information necessary to make climate change spending trade-offs, if in addition to the funding information for science, technology, and international assistance it has previously reported to Congress, OMB also reported similar funding information for programs whose costs were likely to increase due to climate change impacts.

We acknowledge OMB's position that there is significant uncertainty in climate projections, that the ability to distinguish impacts and decisions associated with climate change from other drivers is often not possible, and that any analyses to evaluate those distinctions or assess future trends in expenditures would require substantial resources. The first recommendation does not explicitly call for OMB to provide information on historical or future trends in expenditures related to programs with fiscal exposure to climate change impacts. Additionally, our recommendation does not call for OMB to analyze or distinguish between climate change and other drivers of federal fiscal exposure to climate change impacts, such as coastal development. Instead, our recommendation calls for OMB to provide funding information for federal programs with fiscal exposure to climate change—which could be in the same format as the climate change funding OMB has previously provided to Congress for science, technology, and international assistance. We disagree with OMB's comment that no test exists to identify climate change programs with fiscal exposures and that its identification of such programs would be subjective and not transparent. For example, OMB and others have previously identified federal programs with fiscal exposure to climate change, such as disaster assistance, flood insurance, crop insurance, wildfire management, and other programs that annually incur costs due to weather-related events and that OMB already collects funding information for these programs as part of its preparation of the President's budget request. Further, OMB has previously requested that agencies report information on funds for agency implementation of sustainability goals and resilience measures in the construction and renovation of federal facilities, so it is possible to do so. For these reasons, when the President is required by statute to report on climate change funding, we continue to believe that implementing our recommendation will provide policymakers with more complete information for making important trade-offs between spending with long-term benefits and spending with short-term benefits.

OMB staff also disagreed with our recommendation for OMB to provide Congress with a detailed analysis of federal climate change programs it considers fragmented, overlapping, or duplicative, concurrent with future OMB funding reports. According to the OMB staff, our recommendation is unnecessary because OMB already assesses potential fragmentation, overlap, and duplication as part of its budget process, contained most recently in its Major Savings and Reform volume of the President's budget request. OMB staff also stated that the statutory requirements for the climate change funding reports have been similar year-after-year.

Additionally, the OMB staff said that the United States already prepares an inventory of federal climate-related activities as part of its Climate Action Report to the U.N. Framework Convention on Climate Change. Further, OMB staff also said they are taking our previous recommendations—to develop a federal program inventory to provide greater transparency and consistency in federal program funding and performance information—into account as part of the broader implementation of the Digital Accountability and Transparency Act of

2014 and GRPA Modernization Act of 2010. OMB staff suggested that we revise our second recommendation to direct OMB to more explicitly report on fragmented, overlapping, and duplicative federal climate change programs as part of OMB budget processes.

We agree with OMB that it is important to identify potential fragmented, overlapping, or duplicative programs during the budget process.

However, as we mention in the report, the fiscal year 2018 Major Savings and Reform volume of the budget request generally does not include detailed analyses supporting its justifications for eliminating programs, such as comparisons of similar programs across several characteristics and a comprehensive assessment of potential positive and negative effects of fragmentation, overlap, and duplication, as called for by GAO's criteria. We acknowledge OMB's comment that the statutory requirements for the climate change funding reports have been similar year-after-year. However, we continue to believe that OMB's climate change funding reports do not provide sufficient detail for Congress to identify potential fragmented, overlapping, or duplicative programs, because several agency budget justifications we reviewed included information on climate change activities that were not identifiable based on the level of detail in OMB's funding reports. Further, the 2016 Climate Action Report focuses on progress toward meeting emissions reduction targets and international assistance activities, rather than a comprehensive list of federal climate change programs.

We are encouraged by OMB staff's statement that the agency is taking our previous recommendations on federal program inventories into account as part of its implementation of the Digital Accountability and Transparency Act of 2014 and GRPA Modernization Act of 2010. However, we continue to believe that because OMB collects and reports information on federal climate change funding, OMB is uniquely situated to conduct an assessment of potentially fragmented, overlapping, or duplicative programs across the full range of agencies engaged in climate change activities and that by providing this information to Congress in conjunction with its funding reports, OMB could help decision makers more effectively target limited resources.

As agreed with your office, unless you publicly announce the contents of this chapter earlier, we plan no further distribution until 30 days from the report date. At that time, we will send copies to the appropriate congressional committees, the Director of the Office of Management and Budget, the Director of the Office of Science and Technology Policy, and other interested parties.

Sincerely yours,

J. Alfredo Gómez
Director, Natural Resources and Environment

APPENDIX I: FURTHER INFORMATION ON 18 PRIMARY PURPOSE PROGRAMS IDENTIFIED BY GAO

This appendix presents additional information on primary purpose programs or activities—those described by agencies as having climate change as their primary purpose—as reported by agencies in annual budget justifications.

Climate Research

The Department of Commerce's National Oceanic and Atmospheric Administration's (NOAA) Climate Research program conducts observing and research to predict long-term changes in climate as well as shorter-term variations that are of societal and economic importance.

Program Snapshot

Federal Agency:

Department of Commerce's National Oceanic and Atmospheric Administration (NOAA)

Administering Agency and Relevant Offices:

Climate Program Office (CPO) and Laboratories and Cooperative Institutes within NOAA's Office of Oceanic and Atmospheric Research.

Program Focus Area:

Climate observing.

Grant Eligibility:

Researchers at a variety of institutions and government agencies.

Intended Benefits:

Research intends to benefit communities, decision makers, and other stakeholders' ability to understand and prepare for short and long-term climate variations, such as droughts and natural disasters.

Program Status:

Active.

Sources: GAO analysis of NOAA documents (text); U.S. Government Work (image) | GAO-18-223.

NOAA's Climate Research activities are authorized by the National Climate Program Act.[67]

[67] 15 U.S.C. §§ 2901-2908. The National Climate Program Act requires the President to establish a National Climate Program, with the National Climate Program office within NOAA serving as the lead entity responsible for administering the program

Coordination

NOAA collaborates with other federal agencies through the U.S. Global Change Research Program (USGCRP) and interagency working groups.[68] Further, National Aeronautics and Space Administration (NASA) and NOAA are statutorily required to submit an annual report to relevant congressional committees describing how their earth science programs will be coordinated. [69]

Examples of Program Activities

NOAA's climate research efforts—through its Climate Research Laboratories and Cooperative Institutes; Regional Climate Data and Information; and Climate Competitive Research—include:

- Collecting atmospheric and oceanic observations to understand the state of the oceans and predict climate variability across multiple time- scales. NOAA also uses these data for weather forecasts and warning systems.
- Modeling to understand atmosphere, ocean, biosphere, and cryosphere dynamics to make projections about future marine ecosystems, atmospheric composition, and air quality across multiple time-scales. NOAA's modeling helped develop an improved drought warning system.
- Regional climate information services that provide information and products to improve decision makers' ability to plan for and respond to climate variability and change. NOAA's Regional Integrated Sciences and Assessments program supports regional teams that increase communities' capacity to prepare for and respond to natural disasters.

---

[68] Under the National Science and Technology Council, NOAA is a co-chair of the Committee on Environment, Natural Resources, and Sustainability, and has a leadership role in the following subcommittees: disaster reduction, ocean science and technology, air quality research, and global change research, among others.

[69] 51 U.S.C. § 60505(b).

Atmospheric Systems Research (ASR)

The Department of Energy's (DOE) ASR quantifies the interactions among aerosols, clouds, precipitation, and radiation to improve fundamental understanding of these processes and reduce uncertainty in global and regional climate projections.

Program Snapshot

Federal Agency:
Department of Energy (DOE)
Administering Agency and Relevant Offices:
ASR is in the Climate and Environmental Sciences Division of the Office of Biological and Environmental Research within DOE's Office of Science.
Program focus: Climate modeling. Grant eligibility:
Researchers at a variety of institutions and government agencies, excluding DOE labs, among others.
Intended Benefits:
To improve climate projections and inform the development of solutions to energy and environmental challenges. Other applications include water forecasts.
Program Status:
Active.

Sources: GAO analysis of DOE program documents (text); DOE (image). | GAO-18-223.

According to DOE program documents, the agency created ASR in 2010 by merging two atmospheric research programs that had focused on different measurement scales.[70]

Coordination

ASR coordinates closely with the Atmospheric Radiation Measurement Climate Research Facility program and partners with other Climate and Environmental Sciences Division (CESD) modeling programs.[71] ASR scientists participate in semi-annual meetings to discuss findings and research priorities. At the division level, CESD officials collaborate with other federal agencies through the U.S. Global Change Research Program, and interagency working groups.[72]

Examples of Program Activities

ASR scientists conduct field campaigns and work with the Atmospheric Radiation Measurement Climate Research Facility to conduct research. ASR also supports research through grants.

- Modeling atmospheric processes and developing model parameters that incorporate ASR's research on aerosols, clouds, and precipitation.[73]
- Targeted observing through short-term field campaigns and experiments on specific atmospheric processes to supplement Atmospheric Radiation Measurement Climate Research Facility data.[74] Scientists have used ASR research to improve their ability

---

[70] Specifically, DOE merged its Atmospheric Radiation Measurement Program and its Atmospheric Science Program in 2010 to better utilize the measurement capabilities of the newly expanded Atmospheric Radiation Measurement Climate Research Facility in a cohesive manner.

[71] For example, ASR collaborates with DOE's Regional and Global Climate Modeling, Earth System Modeling, and Terrestrial Ecosystem Science modeling efforts.

[72] Under the National Science and Technology Council, DOE is a member of the Committee on Environment, Natural Resources, and Sustainability, and the Director of CESD participates in the U.S. Group on Earth Observations subcommittee.

[73] DOE, *Accomplishments in Atmospheric Science 2008 – 2013*, DOE-SC-0159.

[74] For example, ASR supplements Atmospheric Radiation Measurement Climate Research Facility's long-term observations with laboratory studies and short-term field campaigns

to predict future changes in precipitation, cloud cover, and atmospheric energy.

Atmospheric Radiation Measurement (ARM) Climate Research Facility

The Department of Energy's (DOE) ARM Climate Research Facility provides researchers with observation facilities to improve the understanding and model representation of clouds and aerosols, as well as their interactions with the Earth's surface.

Program Snapshot

Federal Agency:
Department of Energy (DOE)
Administering Agency and Relevant Offices:
ARM is a scientific user facility within the Climate and Environmental Sciences Division of the Office of Biological and Environmental Research within DOE's Office of Science. An infrastructure board manages the facility, and components of the facility are operated by DOE laboratories.
Program focus: Climate observing. Grant eligibility: Not applicable.
Intended Benefits:
To improve the predictive understanding of Earth's climate, and inform solutions to U.S. energy and environmental challenges. Other applications include storm prediction.
Program Status:
Active.

Sources: GAO analysis of DOE program documents (text); DOE (image). | GAO-18-223.

---

that target specific atmospheric processes in a variety of locations and atmospheric conditions.

According to DOE program documents, the agency established the ARM Climate Research Facility as a national user facility in 2003.[75]

Coordination

ARM coordinates closely with DOE's Atmospheric Systems Research program and partners with other programs within DOE's Climate and Environmental Sciences Division (CESD).[76] ARM also coordinates with federal agencies through its Science Board and constituent groups.[77] At the division level, CESD officials collaborate with other federal agencies through the U.S. Global Change Research Program, and interagency working groups.[78]

Examples of Program Activities

ARM's efforts include:

- Collecting long-term data on aerosols, clouds, and other atmospheric properties from fixed and mobile instruments. Scientists use these data to understand how sunlight, clouds, aerosols, and precipitation affect temperature, weather, and climate.
- High resolution modeling of atmospheric and cloud processes using ARM data. DOE, NOAA, and other scientists use the model outputs to better understand cloud formation, earth system science and severe weather.

---

[75] The observational sites were originally established under DOE's ARM Program, which DOE established in 1990 and merged with its Atmospheric Science Program in 2010 to become the Atmospheric Systems Research program.

[76] For example, ARM collaborates with CESD's Climate and Earth System Modeling, Environmental System Science, and the Environmental Molecular Sciences Laboratory efforts.

[77] For example, the ARM Science Board, which reviews research proposals, includes NASA and DOD officials. Constituent groups such as ARM's User Executive Committee, and Unmanned Aerial Systems Advisory Panel, include representatives from NOAA, the NSF-funded National Center for Atmospheric Research and NASA.

[78] Under the National Science and Technology Council, DOE is a member of the Committee on Environment, Natural Resources, and Sustainability, and the Director of CESD participates in the U.S. Group on Earth Observations subcommittee.

- Providing the use of highly-instrumented in situ and remote sensing observation facilities to researchers. Almost 1200 researchers from academia, government, industry and others used ARM facilities in 2016.[79]

Terrestrial Ecosystem Science (TES)

The Department of Energy's (DOE) TES seeks to improve the representation of terrestrial ecosystem processes in Earth system models to improve climate projections and inform energy decisions.

Program Snapshot

Federal Agency:
Department of Energy (DOE)
Administering Agency and Relevant Offices:
TES is in the Climate and Environmental Sciences Division of the Office of Biological and Environmental Research within DOE's Office of Science.
Program focus: Climate observing. Grant eligibility:
Researchers at a variety of institutions and government agencies.
Intended Benefits:
To improve climate projections and inform DOE's energy decisions.
Program Status:
Active.

Sources: GAO analysis of DOE program documents (text); DOE (image). | GAO-18-223.

[79] DOE/SC-ARM-16-056.

According to DOE program documents, the agency created TES by consolidating predecessor programs on carbon processes and ecological science.[80]

Coordination

TES coordinates closely with DOE's Subsurface Biogeochemical Research program and partners with other programs within DOE's Climate and Environmental Sciences Division and research at DOE scientific facilities.[81] TES also collaborates with other federal agencies through the U.S. Global Change Research Program and interagency working groups.[82]

Examples of Program Activities

TES efforts include:

- Collecting long-term data on carbon flows between land and atmosphere, generally in North America. Scientists at DOE, NOAA, the U.S. Forest Service and others use the data to understand carbon sources and sinks and how they are affected by a variety of factors, such as land-use, disturbance, and climate, among others.[83]
- Using Earth system models to identify research gaps and designing field experiments to address uncertainties. TES provides research results to modeling efforts, to improve climate projections.
- Conducting long-term field experiments focused on regionally or globally significant ecosystems and processes to study their interactions with the climate. Scientists use these data to predict

---

[80] In particular, DOE consolidated its Terrestrial Carbon Processes and Program in Ecological Research programs to become the TES program.

[81] For example, TES and the Subsurface Biogeochemical Research program hold joint investigator meetings annually to discuss ongoing research. TES also coordinates closely with the Climate and Earth System Modeling program, among others.

[82] For example, TES participates in the Carbon Cycle Interagency Working Group that includes representatives from 12 agencies, including NOAA, NASA, USDA, NSF and USGS. As part of the Interagency Arctic Research Policy Committee, TES coordinates its Arctic research with NASA's Arctic research, among others.

[83] DOE, NASA, NOAA, and the U.S. Forest Service supported the formation of the AmeriFlux network—an observation network to measure carbon flows at over 120 sites.

the responses and feedbacks between the climate and these ecosystems, such as permafrost.[84]

Terrestrial Carbon Sequestration Research

The Department of Energy's (DOE) Terrestrial Carbon Sequestration Research supported efforts to identify, understand, and predict the fundamental physical, chemical, biological, and genetic mechanisms controlling carbon sequestration in terrestrial ecosystems.

Program Snapshot

Federal Agency:
Department of Energy (DOE)
Administering Agency and Relevant Offices:
DOE's National Energy Technology Laboratory (NETL) managed the Terrestrial Carbon Sequestration Research program.
Program focus:
Other.
Grant eligibility:
Researchers at a variety of institutions, industry, and DOE labs.
Intended Benefits:
Provide economically competitive and environmentally safe options for offsetting the projected increase in carbon dioxide emissions. Concurrent benefits could include creating wildlife habitat, preventing soil erosion, and boosting local economies, among others.
Program Status:
No longer funded as a separate line item in DOE's budget justification after fiscal year 2011.

Sources: GAO analysis of DOE program documents (text); USDA photo by Lance Cheung (image). | GAO-18-233.

[84] TES is studying ecosystems in the Arctic and in tropical rainforests, and whether their stored carbon will be affected by changing climate conditions.

According to DOE's budget justification, Terrestrial Carbon Sequestration Research first appeared as a separate line item in fiscal year 2010.[85]

Coordination

Terrestrial Carbon Sequestration Research was a joint effort of DOE's Office of Science and DOE's Office of Fossil Energy. At the department level, DOE also coordinated with federal agencies through the Climate Change Technology Program.[86]

Examples of Program Activities

Terrestrial Carbon Sequestration Research efforts included:

- Supporting research to enhance the carbon storage productivity of terrestrial ecosystems, such as by applying additives to the soil or managing plant root systems.
- Supporting research to develop models, methods and tools to evaluate carbon sequestration strategies.
- Fostering partnerships between landowners, government agencies, and energy producers, such as coal and utility companies to help develop the best approaches for increasing carbon sequestration in soils and plants.

Climate and Earth System Modeling (CESM)

The Department of Energy's (DOE) CESM develops model components as well as Earth system models that include human impacts and natural systems to simulate climate variability (e.g., El Niño) and predict change from decades to centuries at regional and global scales.

---

[85] However, DOE conducted carbon sequestration research as part of other programs prior to 2010 as well as after DOE stopped listing Terrestrial Carbon Sequestration Research as a separate line item in its budget justifications

[86] According to DOE program documents, the Climate Change Technology Program was created to coordinate and prioritize the federal government's investments in climate related technology, research, development, demonstration, and deployment.

Program Snapshot

Federal Agency:
National Aeronautics and Space Administration (NASA)
Administering Agency and Relevant Offices:
CPF is a component of the CLARREO mission within NASA's Earth Systematic Missions, in its Earth Science Division, which is part of NASA's Science Mission Directorate. NASA's Langley Research Center leads the CLARREO mission.
Program focus: Climate observing. Grant eligibility: Not applicable.
Intended Benefits:
CPF will reduce risks for the CLARREO mission by demonstrating calibration approaches with other satellite instruments. CPF's establishment of a calibration standard for other orbiting sensors will also improve weather forecasting and the accuracy of climate observations.
Program Status:
Projected launch no earlier than 2022.

Sources: GAO analysis of NASA program documents (text); NASA (image). | GAO-18-223.

NASA's Earth Science program was established in response to a requirement in the National Aeronautics and Space Administration Authorization Act, Fiscal Year 1993.[95]

[95] Pub. L. No. 102-588, tit. I, § 102(g) (Nov. 4, 1992) (requiring the NASA Administrator to carry out an Earth Observing System program that addresses the highest priority international climate change research goals).

Coordination

The CLARREO team coordinates with NASA's Goddard Space Flight Center and Jet Propulsion Laboratory. The CLARREO team also coordinates with the National Oceanic and Atmospheric Administration (NOAA), the National Institute for Standards and Technology (NIST), the Department of Energy (DOE), and academic partners.[96] At the division level, NASA Earth Science collaborates directly with NOAA, and with other federal agencies through the U.S. Global Change Research Program and interagency working groups.[97]

Examples of Program Activities

CPF Mission efforts include:

- Develop and demonstrate one highly accurate spectrometer instrument to measure reflected solar energy and to serve as an in-orbit standard to calibrate and improve the accuracy of other satellite instruments.
- One year of operation on the International Space Station, and one year of data analysis.

Glory Mission

The National Aeronautics and Space Administration's (NASA) Glory Mission's goal was to examine how aerosols and solar energy affect the

[96] For example, the full CLARREO mission includes inter-calibration and comparison software developed jointly by NASA and NOAA. The NASA and NOAA Administrators are required to review and monitor, through a joint working group, the agencies' missions to ensure maximum coordination in the design, operation, and transition of missions, where appropriate, and submit annual reports to relevant congressional committees on how their earth science programs will be coordinated. 51 U.S.C. §§ 60505(a),(b).

[97] For example, under the NTSC, NASA is a member of the Committee on Environment, Natural Resources, and Sustainability, and the former Council on Climate Preparedness and Resilience. NASA also has a leadership role in the Working Group on Ocean Acidification. NASA coordinates with international and commercial partners through its data system communities, such as the Committee on Earth Observation Satellites.

climate, by collecting data on atmospheric aerosol properties and incoming solar energy.

The Glory Mission was part of NASA's Earth Science program, which was established in response to a requirement in the National Aeronautics and Space Administration Authorization Act, Fiscal Year 1993.[98] The Glory Mission was reauthorized by the National Aeronautics and Space Administration Authorization Act of 2008.[99]

Program Snapshot

Federal Agency:
National Aeronautics and Space Administration (NASA)
Administering Agency and Relevant Offices:
Glory was located in NASA's Earth Systematic Missions, in its Earth Science Division, which is part of NASA's Science Mission Directorate. NASA's Goddard Space Flight Center led the Glory Mission.
Program focus: Climate observing. Grant eligibility: Not applicable.
Intended Benefits:
Glory would have increased the accuracy of aerosol and solar energy measurements to improve understanding of the Earth's energy budget and climate change regional impacts.
Program Status:
No longer funded as a separate line item in NASA's budget justification. The satellite failed to reach orbit in fiscal year 2011.

Sources: GAO analysis of NASA program documents (text); NASA (image). | GAO-18-223.

[98] Pub. L. No. 102-588, tit. I, § 102(g) (Nov. 4, 1992).
[99] Pub. L. No. 110-422, tit. II, § 206(a) (Oct. 15, 2008).

Coordination

The Glory Mission team coordinated closely with NASA's Goddard Institute for Space Studies. The Glory team also coordinated with the National Oceanic and Atmospheric Administration (NOAA), the National Institute for Standards and Technology (NIST), the Naval Research Laboratory, and other partners.[100] At the division level, NASA Earth Science collaborates directly with NOAA, and with other federal agencies through the U.S. Global Change Research Program and interagency working groups.[101]

Examples of Program Activities

Glory Mission efforts included:

- Developing and testing two instruments, the Aerosol Polarimetry Sensor and the Total Irradiance Monitor, which were intended to supply highly accurate measurements of aerosol properties and incoming solar radiation.
- The Glory Mission spacecraft launched in March 2011, but failed to reach orbit due to a launch vehicle equipment failure.[102]

The Ice, Cloud, and Land Elevation Satellite (ICESat)

The National Aeronautics and Space Administration's (NASA) ICESat measured changes in ice sheet mass and elevation to understand how changing conditions affect ice and sea level.

---

[100] For example, the NASA and NOAA Administrators are required to review and monitor, through a joint working group, the agencies' missions to ensure maximum coordination in the design, operation, and transition of missions, where appropriate, and submit annual reports to relevant congressional committees on how their earth science programs will be coordinated. 51 U.S.C. §§ 60505(a),(b).

[101] For example, under the NTSC, NASA is a member of the Committee on Environment, Natural Resources, and Sustainability, and the former Council on Climate Preparedness and Resilience. NASA also has a leadership role in the Working Group on Ocean Acidification. NASA coordinates with international and commercial partners through its data system communities, such as the Committee on Earth Observation Satellites.

[102] NASA, *Overview of the Glory Mishap Investigation Results for Public Release.*

Program Snapshot

Federal Agency:
National Aeronautics and Space Administration (NASA)
Administering Agency and Relevant Offices:
ICESat was located in NASA's Earth Systematic Missions, in its Earth Science Division, which is part of NASA's Science Mission Directorate. NASA's Goddard Space Flight Center led the ICESat mission.
Program focus: Climate observing.
Grant eligibility: Not applicable Intended Benefits:
ICESat provided data on ice sheet changes to better understand the climate system.
Program Status:
No longer funded as a separate line item in NASA's budget justification after completing its mission in fiscal year 2010.

Sources: GAO analysis of NASA program documents (text); NASA (image). | GAO-18-223.

ICESat is part of NASA's Earth Science program, which was established in response to a requirement in the National Aeronautics and Space Administration Authorization Act, Fiscal Year 1993.[103]

Coordination

The ICESat mission team coordinated with industry and academic partners.[104] At the division level, NASA Earth Science collaborates directly

[103] Pub. L. No. 102-588, tit. I, § 102(g) (Nov. 4, 1992).
[104] NASA, Goddard Space Flight Center, *ICESat: Ice, Cloud, and Land Elevation Satellite,* FS-2002-9-047-GSFC.

with the National Oceanic and Atmospheric Administration (NOAA), and with other federal agencies through the U.S. Global Change Research Program and interagency working groups.[105]

Examples of Program Activities

ICESat mission efforts included:

- Developing and launching one instrument, the Geoscience Laser Altimeter System, which quantified ice sheet mass balance and measured global distributions of clouds and aerosols, topography, sea ice and vegetation cover.
- ICESat data led to scientific advances in measuring changes in the mass of the Greenland and Antarctic ice sheets and polar sea ice thickness.[106] It also led to detailed records of land elevation, tree canopies, global standing carbon stocks and terrestrial water storage capacity.
- ICESat completed its mission goal to provide detailed ice elevation data in 2008, and ceased operation in fiscal year 2010.[107]

The Ice, Cloud, and Land Elevation Satellite-2 (ICESat-2)

The National Aeronautics and Space Administration's (NASA) ICESat-2 will continue ICESat's mission to measure changes in ice sheet

[105] For example, under the NTSC, NASA is a member of the Committee on Environment, Natural Resources, and Sustainability, and the former Council on Climate Preparedness and Resilience. NASA also has a leadership role in the Working Group on Ocean Acidification. NASA coordinates with international and commercial partners through its data system communities, such as the Committee on Earth Observation Satellites.

[106] According to NASA, the Geoscience Laser Altimeter System instrument allowed scientists to measure the multi-year height of atmospheric cloud and aerosol layers for the first time to improve understanding of the Earth's climate. For a list of publications that used ICESat data, see http://nsidc.org/data/icesat/research.html.

[107] ICESat operated from 2003 to 2009 and completed its original mission goal in 2008. NP-2010-7-160-GSFC and NASA, *ICESat: Ice, Cloud, and Land Elevation Satellite 2 (ICESat-2) Application Plan.*

mass and elevation to understand how changing conditions affect ice and sea level.

Program Snapshot

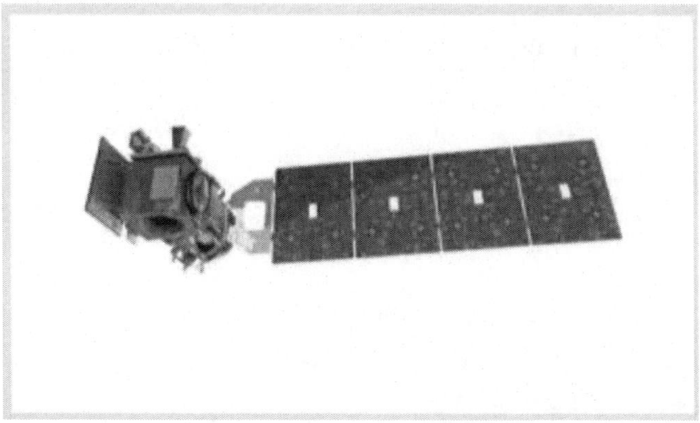

Federal Agency:
National Aeronautics and Space Administration (NASA)
Administering Agency and Relevant Offices:
ICESat-2 is located in NASA's Earth Systematic Missions, in its Earth Science Division, which is part of NASA's Science Mission Directorate. NASA's Goddard Space Flight Center leads the ICESat-2 mission.
Program focus: Climate observing.
Grant eligibility: Not applicable.
Intended Benefits:
To improve the measurement precision for polar ice sheet mass, sea ice, and other elevation data to better understand the Earth's climate system.
Program Status:
Projected launch in 2018.

Sources: GAO analysis of NASA program documents (text); Orbital Sciences Corporation (image). | GAO-18-223.

ICESat-2 is part of NASA's Earth Science program, which was established in response to a requirement in the National Aeronautics and Space Administration Authorization Act, Fiscal Year 1993.[108]

[108] Pub. L. No. 102-588, tit. I, § 102(g) (Nov. 4, 1992).

Coordination

The ICESat-2 team coordinates with the U.S. Geological Survey, National Oceanic and Atmospheric Administration (NOAA), academic partners, and others.[109] At the division level, NASA Earth Science collaborates directly with NOAA, and with other federal agencies through the U.S. Global Change Research Program and interagency working groups.[110]

Examples of Program Activities

ICESat-2 mission efforts include:

- Developing, testing, and launching one instrument, the Advanced Topographic Laser Altimeter System (ATLAS) to provide high-resolution elevation data.[111]
- IceSat-2 will measure changes in polar ice sheet mass, sea ice thickness, and vegetation-canopy heights. It will also provide atmospheric profiles, ocean elevation, and inland waterelevations.
- Other data applications may include improving the Navy's ice forecasting system and improving global forecasting for floods and droughts.
- Data will be sent to the National Snow and Ice Data Center, which is jointly supported by NASA, NOAA, and the National Science Foundation.

[109] Scientists from NASA's Jet Propulsion Laboratory also assisted with IceSat-2.

[110] For example, under the NTSC, NASA is a member of the Committee on Environment, Natural Resources, and Sustainability, and the former Council on Climate Preparedness and Resilience. NASA also has a leadership role in the Working Group on Ocean Acidification. NASA coordinates with international and commercial partners through its data system communities, such as the Committee on Earth Observation Satellites.

[111] We previously reported that ICESat-2 has encountered problems with cracked crystals in the flight lasers in ATLAS that will likely cause it to miss its committed launch date and could cause it to exceed its current cost baseline. The project was previously re-baselined in 2014 because of development and design issues, and now plans to launch no earlier than September 2018—11 months later than its previously planned October 2017 launch date. GAO, *NASA: Assessment of Major Projects,* GAO-17-303SP (Washington, D.C.: May 17, 2016.)

Orbiting Carbon Observatory-2 (OCO-2)

The National Aeronautics and Space Administration's (NASA) OCO-2 measures carbon dioxide concentrations, to better understand the carbon cycle.

Program Snapshot

Federal Agency:
National Aeronautics and Space Administration (NASA)
Administering Agency and Relevant Offices:
OCO-2 is located in NASA's Earth System Science Pathfinder program in its Earth Science Division, which is part of NASA's Science Mission Directorate. NASA's Jet Propulsion Laboratory leads the OCO-2 mission.
Program focus: Climate observing.
Grant eligibility: Not applicable.
Intended Benefits:
To improve global carbon dioxide measurements to better understand the carbon cycle.
Program Status:
Operational since 2014.

Sources: GAO analysis of NASA program documents (text); NASA (image). | GAO-18-223.

OCO-2 is part of NASA's Earth Science program, which was established in response to a requirement in the National Aeronautics and

Space Administration Authorization Act, Fiscal Year 1993.[112] The National Aeronautics and Space Administration Authorization Act of 2000 required NASA to develop a carbon cycle remote sensing applications research program.[113]

Coordination

The OCO-2 team coordinated with NASA's Goddard Space Flight Center and Langley Research Center. The team also coordinated with the Department of Energy, National Oceanic and Atmospheric Administration (NOAA) and others.[114] At the division level, NASA Earth Science collaborates directly with NOAA, and with other federal agencies through the U.S. Global Change Research Program and interagency working groups.[115]

Examples of Program Activities

OCO-2 mission efforts include:

- Developed and launched one instrument with three spectrometers to provide high-resolution measurements of carbon dioxide concentrations globally.
- OCO-2 measures carbon dioxide concentrations and its geographic distribution to improve understanding of the global carbon cycle and predictions of global climate change. Other applications may include evaluating carbon cycle impacts for land use and energy policy options.[116]

---

[112] Pub. L. No. 102-588, tit. I, § 102(g) (Nov. 4, 1992).

[113] Pub. L. No. 106-391, tit. III, § 315(a)(1) (Oct. 30, 2000).

[114] The OCO team worked with international research institutions and private sector partners for the original mission that failed during launch in 2009. NASA reformulated the OCO science team to continue their work for the follow-on mission, OCO-2.

[115] For example, under the NTSC, NASA is a member of the Committee on Environment, Natural Resources, and Sustainability, and the former Council on Climate Preparedness and Resilience. NASA also has a leadership role in the Working Group on Ocean Acidification. NASA coordinates with international and commercial partners through its data system communities, such as the Committee on Earth Observation Satellites.

[116] NASA, *Orbiting Carbon Observatory-2 Launch Press Kit* (July 2014).

Orbiting Carbon Observatory-3 (OCO-3)

The National Aeronautics and Space Administration's (NASA) OCO-3 will continue OCO-2's mission measuring carbon dioxide concentrations, to better understand the carbon cycle.

Program Snapshot

Federal Agency:
National Aeronautics and Space Administration (NASA)
Administering Agency and Relevant Offices:
OCO-3 is located in NASA's Earth System Science Pathfinder program in its Earth Science Division, which is part of NASA's Science Mission Directorate. NASA's Jet Propulsion Laboratory leads the OCO-3 mission.
Program focus: Climate observing.
Grant eligibility: Not applicable.
Intended Benefits:
To improve carbon dioxide measurements to better understand the carbon cycle.
Program Status:
Launch date to be determined.

Sources: GAO analysis of NASA program documents (text); NASA (image). | GAO-18-223.

OCO-3 is part of NASA's Earth Science program, which was required established in response to a requirement in the National Aeronautics and

Space Administration Authorization Act, Fiscal Year 1993.[117] The National Aeronautics and Space Administration Authorization Act of 2000 required NASA to develop a carbon cycle remote sensing applications research program.[118]

Coordination

The OCO-3 team coordinated with NASA's Human Exploration and Operations Mission Directorate, among others. The team also coordinated with the Department of Energy, the National Oceanic and Atmospheric Administration (NOAA) and others.[119] At the division level, NASA Earth Science collaborates directly with NOAA, and with other federal agencies through the U.S. Global Change Research Program and interagency working groups.[120]

Examples of Program Activities

OCO-3 mission efforts include:

- Developing and launching one spectrometry instrument to the International Space Station to provide high-resolution measurements of carbon dioxide and chlorophyll fluorescence globally.
- OCO-3 will use an improved sampling strategy to provide highly accurate measurements of carbon dioxide concentrations, including the ability to explore daily variations in carbon dioxide at a

---

[117] Pub. L. No. 102-588, tit. I, § 102(g) (Nov. 4, 1992).

[118] Pub. L. No. 106-391, tit. III, § 315(a)(1) (Oct. 30, 2000).

[119] NASA worked with international research institutions and private sector partners for the original OCO mission that failed in 2009. The OCO-3 mission will use the OCO-2 mission's spare spectrometer, which was based on the original OCO design.

[120] For example, under the NTSC, NASA is a member of the Committee on Environment, Natural Resources, and Sustainability, and the former Council on Climate Preparedness and Resilience. NASA also has a leadership role in the Working Group on Ocean Acidification. NASA coordinates with international and commercial partners through its data system communities, such as the Committee on Earth Observation Satellites.

regional scale—as they relate to changes in urban population and fossil fuel use—to improve understanding of the carbon cycle.[121]

Solar Radiation and Climate Experiment (SORCE)

The National Aeronautics and Space Administration's (NASA) SORCE measures solar radiation properties to better understand solar variability and its effects.

Program Snapshot

Federal Agency:
National Aeronautics and Space Administration (NASA)
Administering Agency and Relevant Offices:
SORCE is located in NASA's Earth Systematic Missions in its Earth Science Division, which is part of NASA's Science Mission Directorate. NASA's Goddard Space Flight Center provides scientific and management oversight and support.
Program focus: Climate observing.
Grant eligibility: Not applicable.
Intended Benefits:
To measure solar radiation to better understand solar variability and its climate effects.
Program Status:
Operational since 2003.

Sources: GAO analysis of NASA program documents (text); NASA (image). | GAO-18-223.

[121] Eldering et. al., *The OCO-3 Mission: Overview of Science Objectives and Status*, Geophysical Research Abstracts, Vol. 18, EGU2016-5189 (2016).

SORCE is part of NASA's Earth Science program, which was established in response to a requirement in the National Aeronautics and Space Administration Authorization Act, Fiscal Year 1993.[122]

Coordination

SORCE is a joint partnership between NASA and the University of Colorado's Laboratory for Atmospheric and Space Physics (LASP) in Boulder, Colorado. LASP designed, built, and tested the scientific instruments. LASP also processes and distributes SORCE data.[123] At the division level, NASA Earth Science collaborates directly with the National Oceanic and Atmospheric Administration (NOAA), and with other federal agencies through the U.S. Global Change Research Program and interagency working groups.[124]

Examples of Program Activities

SORCE mission efforts include:

- Developing and launching a satellite with four instruments to measure the wavelengths of incoming solar radiation (e.g., x-ray, ultraviolet, visible, and near-infrared) and total solar radiation.
- Improving the precision of previous solar radiation measurements and continuing the long-term data record of solar observations.[125]

Provided data to help estimate past and future solar behavior and climate response and inform research on long-term climate change, natural

---

[122] Pub. L. No. 102-588, tit. I, § 102(g) (Nov. 4, 1992).

[123] Goddard Space Flight Center, *NASA Facts: Solar Radiation and Climate Experiment (SORCE)*, FS-2002-12-052-GSFC. Goddard Earth Sciences Data and Information Services Center archives SORCE data.

[124] For example, under the NTSC, NASA is a member of the Committee on Environment, Natural Resources, and Sustainability, and the former Council on Climate Preparedness and Resilience. NASA also has a leadership role in the Working Group on Ocean Acidification. NASA coordinates with international and commercial partners through its data system communities, such as the Committee on Earth Observation Satellites.

[125] SORCE was designed to operate from 2003 to 2009, but has continued operating beyond its planned lifespan.

variability and enhanced climate prediction, as well as atmospheric ozone and ultraviolet radiation.

Total Solar Irradiance Sensor-1 (TSIS-1)

The National Aeronautics and Space Administration's (NASA) TSIS-1 continues SORCE's mission measuring solar radiation properties to better understand solar variability and its effects.

Program Snapshot

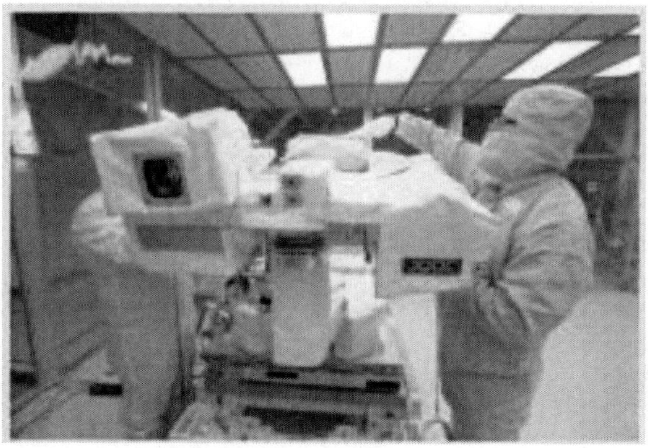

Federal Agency:
National Aeronautics and Space Administration (NASA)
Administering Agency and Relevant Offices:
TSIS-1 is located in NASA's Earth Systematic Missions in its Earth Science Division, which is part of NASA's Science Mission Directorate. NASA's Goddard Space Flight Center manages and oversees the project.
Program focus: Climate observing.
Grant eligibility: Not applicable.
Intended Benefits:
To measure solar radiation to understand solar variability and its climate effects.
Program Status:
Launched successfully in December 2017. Instruments powered on in March 2018.

Sources: GAO analysis of NASA program documents (text); NASA (image). | GAO-18-223.

TSIS-1 is part of NASA's Earth Science program, which was established in response to a requirement in the National Aeronautics and Space Administration Authorization Act, Fiscal Year 1993.[126]

Coordination

TSIS-1 is a joint partnership between NASA and the University of Colorado's Laboratory for Atmospheric and Space Physics (LASP) in Boulder, Colorado. LASP designed, built, and tested the instruments, and will process the data. At the division level, NASA Earth Science collaborates directly with the National Oceanic and Atmospheric Administration (NOAA), and with other federal agencies through the U.S. Global Change Research Program and interagency working groups.[127]

Examples of Program Activities

TSIS-1 mission efforts include:

- Developing and launching two instruments to the International Space Station to measure incoming solar radiation properties and total solar radiation.
- Improving the precision of SORCE's solar radiation measurements and continuing the long-term data record of solar observations.[128]
- Data can be used to study the sun's influence on the ozone layer, atmospheric circulation, clouds, and ecosystems. These measurements are also critical to understanding current and future climate conditions.[129]

---

[126] Pub. L. No. 102-588, tit. I, § 102(g) (Nov. 4, 1992).

[127] For example, under the NTSC, NASA is a member of the Committee on Environment, Natural Resources, and Sustainability, and the former Council on Climate Preparedness and Resilience. NASA also has a leadership role in the Working Group on Ocean Acidification. NASA coordinates with international and commercial partners through its data system communities, such as the Committee on Earth Observation Satellites.

[128] To bridge the potential data gap between the aging SORCE satellite and the TSIS-1 mission, NASA developed the Total Solar Irradiance Calibration Transfer Experiment using spare parts, which was launched in 2013 as part of the Air Force Space Test Program Satellite, to help calibrate measurements between SORCE and TSIS-1.

[129] NASA, *TSIS-1: Measuring the Sun's Energy Input to Earth,* accessed at https://www.nasa.gov/sites/default/files/atoms/files/tsis_fact_sheet-042617-v1.pdf.

Total Solar Irradiance Sensor-2 (TSIS-2)

The National Aeronautics and Space Administration's (NASA) TSIS-2 will continue TSIS-1's mission measuring solar radiation properties to better understand solar variability and its effects.

Program Snapshot

Federal Agency:
National Aeronautics and Space Administration (NASA)
Administering Agency and Relevant Offices:
TSIS-2 is located in NASA's Earth Systematic Missions in its Earth Science Division, which is part of NASA's Science Mission Directorate.
Program focus: Climate observing.
Grant eligibility: Not applicable.
Intended Benefits:
To measure solar radiation to understand solar variability and its climate effects.
Program Status:
Projected launch in 2022.

Sources: GAO analysis of NASA program documents (text); NASA (image). | GAO-18-223.

TSIS-2 is part of NASA's Earth Science program, which was established in response to a requirement in the National Aeronautics and Space Administration Authorization Act, Fiscal Year 1993.[130]

[130] Pub. L. No. 102-588, tit. I, § 102(g) (Nov. 4, 1992).

Coordination

NASA has contracted with the University of Colorado's Laboratory for Atmospheric and Space Physics (LASP) in Boulder, Colorado for a study on two compact instruments for TSIS-2. At the division level, NASA Earth Science collaborates directly with the National Oceanic and Atmospheric Administration (NOAA), and with other federal agencies through the U.S. Global Change Research Program and interagency working groups.[131]

Examples of Program Activities

TSIS-2 mission efforts will include:

- Developing and launching two instruments to the International Space Station to measure incoming solar radiation properties and total solar radiation.
- Sustain TSIS-1's mission to continue the long-term data record of solar observations.
- Data may be used to study the sun's influence on the ozone layer, atmospheric circulation, clouds, and ecosystems. These measurements are also critical to understanding current and future climate conditions.[132]

Agroclimatology

The U.S. Department of Agriculture's (USDA) Agroclimatology (formerly Global Change, Ultraviolet Monitoring and Research Program) supports an ultraviolet radiation observing network, and uses the data to inform climate forecasting models and assess ultraviolet radiation effects.

---

131 For example, under the NTSC, NASA is a member of the Committee on Environment, Natural Resources, and Sustainability, and the former Council on Climate Preparedness and Resilience. NASA also has a leadership role in the Working Group on Ocean Acidification. NASA coordinates with international and commercial partners through its data system communities, such as the Committee on Earth Observation Satellites.

132 NASA, *TSIS-1: Measuring the Sun's Energy Input to Earth,* accessed at https://www.nasa.gov/sites/default/files/atoms/files/tsis_fact_sheet-042617-v1.pdf.

Program Snapshot

Federal Agency:
U.S. Department of Agriculture's National Institute for Food and Agriculture (NIFA)
Administering Agency and Relevant Offices:
Colorado State University's Natural Resource Ecology Laboratory administers the program. NIFA's Division of Agricultural Systems, within its Institute of Food Production and Sustainability, provides grant funding and oversight.
Program focus: Climate observing.
Grant eligibility: Land-grant institutions, among others.
Intended Benefits:
To improve understanding of ultraviolet radiation impacts and climate change on agriculture. Other applications include atmospheric, biological, ecological, human health and materials sciences.
Program Status:
Active.

Sources: GAO analysis of USDA program documents (text); USDA photo by Jack Dykinga (image). | GAO-18-223.

USDA established the Ultraviolet Monitoring and Research Program in 1992 pursuant to the Competitive, Special, and Facilities Research Grant Act.[133]

Coordination

The program is affiliated with other academic facilities, including the joint NASA – University of Maryland Earth System Science

[133] Pub. L. No. 89-106, § 2 (Aug. 4, 1965) (*codified as amended at* 7 U.S.C. § 3157). According to agency documents, USDA established the Ultraviolet Monitoring and Research Program to measure ultraviolet radiation levels over wide geographic areas of the United States and to assess the impact of ultraviolet radiation on crops, plants, animals, and ecosystems.

Interdisciplinary Center.[134] USDA's National Institute for Food and Agriculture (NIFA) coordinates with several other USDA agencies, such as the Agricultural Research Service.[135] NIFA also collaborates with other federal agencies, contributes to the U.S. Global Change Research Program, and participates in interagency working groups.[136]

Examples of Program Activities

Program efforts include:

- Observing ultraviolet radiation and photosynthetically-active radiation through a network of 38 sites that provide the only source of long-term surface measurements of ultraviolet radiation.[137]
- Studying ultraviolet radiation effects and assessing the combined effect of ultraviolet radiation with other climate stress factors, such as moisture, temperature, and carbon dioxide.
- Developing modeling tools that integrate observing data, ultraviolet effects studies, and satellite observations to study how climate and crop production interact and associated effects on management practices and agricultural economics.

Climate Change Program Office (CCPO)

The U.S. Department of Agriculture's (USDA) CCPO coordinates agricultural, rural and forestry-related climate change program and policy issues department-wide.

---

[134] The program is also affiliated with the Soil-Plant-Atmosphere-Research Facility at Mississippi State University.

[135] NIFA coordinates with USDA's Forest Service and USDA's Natural Resources Conservation Service, among others.

[136] For example, NIFA collaborates with Commerce, DOD, DOE, Interior, EPA, and NSF, among others. Under the interagency National Science and Technology Council, USDA is a member of the Committee on Environment, Natural Resources, and Sustainability.

[137] Data users have included researchers from about 134 academic institutions, 51 state and federal agencies, 15 international agencies, and 42 commercial enterprises, among others.

Program Snapshot

Federal Agency:
U.S. Department of Agriculture (USDA)
Administering Agency and Relevant Offices:
CCPO is located within USDA's Office of the Chief Economist, in USDA's Office of the Secretary.
Program focus: Climate adaptation.
Grant eligibility: Not applicable.
Intended Benefits:
To provide leadership, support and coordination across USDA agencies in helping the US agriculture and forestry sectors assess and manage greenhouse gas emissions as well as prepare for and adapt to climate change.
Program Status:
Active.

Sources: GAO analysis of USDA program documents (text); USDA photo by Jack Dykinga (image). | GAO-18-223.

USDA created the CCPO in 2010 within the Office of the Chief Economist.[138]

Coordination

CCPO coordinates climate change program and policy issues department-wide.[139] CCPO is USDA's focal point for climate change with

[138] Prior to 2010, USDA's Global Change Program Office facilitated the agency's participation in the U.S. Climate Change Science Program and the U.S. Climate Change Technology Program, according to USDA's fiscal year 2010 budget justification.

[139] For example, CCPO's Director served as the Chair of USDA's Global Change Task Force, which included representatives from twenty USDA agencies and offices.

other federal agencies, the U.S. Global Change Research Program, and interagency working groups.[140]

Examples of Program Activities

CCPO's efforts include:

- Leading USDA adaptation planning efforts to identify how climate change may affect USDA's mission and to ensure adaptation is integrated into USDA programs, policies and operations.
- Publishing the greenhouse gas inventory for agriculture and forestry, in conjunction with other USDA and external partners.
- Representing USDA and coordinating USDA participation and technical support for interagency working groups and international meetings.
- Providing data, tools and information to assist land managers, stakeholders and USDA agencies to prepare for relevant climate change impacts, such as changes in water availability and wildfires, with adaptation assessments, planning and implementation, as well as guidance for measuring greenhouse gas emissions.

APPENDIX II: OMB-REPORTED CLIMATE CHANGE FUNDING SINCE 1993

As reported by OMB, federal funding for climate change research, clean energy technology, international assistance, and adaptation in nominal dollars has increased from about $2.4 billion in fiscal year 1993 to about $13.2 billion in proposed budget authority for fiscal year 2017, with an additional $26.1 billion for climate change programs and activities provided by the American Recovery and Reinvestment Act of 2009.[141] As

[140] Under the interagency National Science and Technology Council, USDA is a member of the Committee on Environment, Natural Resources, and Sustainability, among others.

[141] When adjusted for inflation, reported climate change funding increased from $3.6 billion in 1993 to $12.9 billion in the proposed budget authority for 2017.

shown in Figure 2, the Office of Management and Budget (OMB) has reported federal climate change funding in three main categories since 1993—clean energy technology to reduce emissions, science to better understand climate change, and international assistance for adaptation, clean energy, and sustainable landscapes.

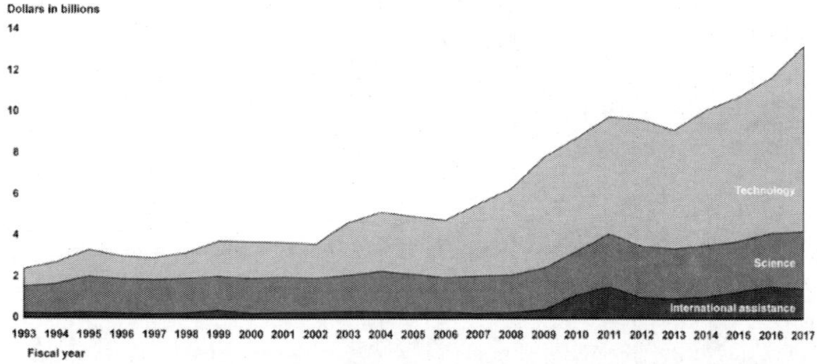

Note: The figure presents actual budget authority except for fiscal years 2011, 2013, and 2017. For fiscal years 2011 and 2017, OMB reported proposed budget authority. For fiscal year 2013, OMB reported final operating level funding. Funding is reported in nominal dollars, which are not adjusted for inflation.

Source: GAO analysis of Office of Management of Budget (OMB) reports. | GAO-18-223.

Figure 2. OMB-Reported Federal Climate Change Funding by Category, from Fiscal Years 1993 through 2017.

OMB has also reported on federal funding for wildlife and natural resource adaptation from fiscal years 2010 through 2013. However, the data OMB reported in the adaptation category does not fully represent adaptation funding as it only includes data from the Department of the Interior. OMB reports Department of the Interior funding for adaptation as follows: fiscal year 2010 $65 million, fiscal year 2011 $87 million, fiscal year 2012 $88 million. We did not include funding for this category in 2013 because we used OMB's fiscal year 2017 report for 2013 funding, and OMB did not report funding under this category in its 2017 report.

Funding in the American Recovery and Reinvestment Act of 2009 (Pub.L. No. 111-5 (Feb. 17, 2009)) and tax expenditures related to climate change are not included in this figure to maintain comparability with earlier GAO reports.

APPENDIX III: CLIMATE CHANGE FUNDING BY CATEGORY AS REPORTED BY THE OFFICE OF MANAGEMENT AND BUDGET (OMB) FROM FISCAL YEARS 1993 THROUGH 2017 (DOLLARS IN MILLIONS)

Funding Category	**1993**	**1994**	**1995**	**1996**	**1997**	**1998**	**1999**	**2000**	**2001**	**2002**	**2003**	**2004**	**2005**	**2006**	**2007**	**2008**	**2009**	**2010**	**2011**	**2012**	**2013**	**2014**	**2015**	**2016**	**2017**
Technology	845	1,038	1,283	1,106	1,056	1,251	1,694	1,793	1,675	1,637	2,555	2,868	2,808	2,789	3,485	4,196	5,386	5,504	5,690	6,121	5,755	6,567	6,999	7,539	8,978
Science	1,306	1,444	1,760	1,654	1,656	1,677	1,657	1,687	1,728	1,667	1,766	1,976	1,864	1,691	1,825	1,832	2,023	2,122	2,561	2,506	2,417	2,492	2,461	2,584	2,775
International Assistance[a]	201	186	228	192	164	186	325	177	218	224	270	252	234	249	188	227	373	1,080	1,494	958	916	1,013	1,252	1,527	1,426
Natural Resources Adaptation[b]	—	—	—	—	—	—	—	—	—	—	—	—	—	—	—	—	—	65	87	88	—	—	—	—	—
Adjustments for programs included in multiple categories	—	—	—	—	—	—	—	—	—	—	—	—	—	—	—	—	—	—	—	*(24)*	*(26)*	*(27)*	*(23)*	*(23)*	*(26)*
Total	**2,352**	**2,668**	**3,271**	**2,952**	**2,876**	**3,114**	**3,535**	**3,511**	**3,603**	**3,522**	**4,584**	**5,090**	**5,269**	**5,876**	**5,498**	**6,255**	**7,782**	**8,771**	**9,832**	**9,649**	**9,062**	**10,045**	**10,689**	**11,627**	**13,153**

Source: GAO analysis of Office of Management and Budget (OMB) reports. | GAO-18-223.

Notes: Dashes indicate that OMB did not report a value for the account for that year. Totals may not add due to rounding.

The table presents actual budget authority except for fiscal years 2011, 2013, and 2017. For fiscal years 2011 and 2017, OMB reported proposed budget authority. For fiscal year 2013, OMB reported final operating level funding. Funding is reported in nominal dollars, which are not adjusted for inflation.

Funding from the American Recovery and Reinvestment Act of 2009 (Pub. L. No. 111-5 (Feb. 17, 2009)) and tax expenditures related to climate change are not included in this table to maintain comparability with earlier GAO reports.

[a]International Assistance includes congressionally appropriated assistance by core agencies (i.e., Department of State, Department of Treasury, U.S. Agency for International Development), as well as complementary agencies (e.g., Environmental Protection Agency), but does not include indirect climate assistance nor development finance and export credit agencies. Additionally, the International Assistance total contains funds that are also counted in the Science and Technology totals. Table total line excludes this double count.

[b]In fiscal year 2010 through 2013, OMB reported the Department of the Interior funding for adaptation under interim categories called "wildlife adaptation" and "natural resources adaptation," while the administration developed criteria to systematically account for a broader suite of adaptation programs. However, the data OMB reported in the adaptation category does not fully represent adaptation as it only includes data from the Department of the Interior. We did not include funding for this category in 2013 because we used OMB's fiscal year 2017 report for 2013 funding, and OMB did not report funding under this category in its 2017 report.

Appendix IV: Climate Change Funding by Agency as Reported by the Office and Management and Budget (OMB), from Fiscal Years 1993 through 2017 (Dollars in Millions)

Agency	1993	1994	1995	1996	1997	1998	1999	2000	2001	2002	2003	2004	2005	2006	2007	2008	2009	2010	2011	2012	2013	2014	2015	2016	2017
Department of Agriculture	55	56	60	52	57	53	138	132	54	59	104	116	110	110	109	116	322	567	426	394	274	517	418	405	446
Department of Commerce	66	63	120	113	102	89	93	91	93	100	156	144	146	253	258	286	392	378	455	359	349	384	367	337	397
Department of Defense	—	—	—	—	—	—	—	—	—	—	83	51	59	77	101	176	261	226	143	481	458	469	486	618	643
Department of Energy	963	1,113	1,173	1,008	968	1,186	1,536	1,652	1,665	1,636	2,214	2,519	2,469	2,504	3,158	3,791	4,711	4,564	5,074	4,599	4,404	4,931	5,410	5,844	7,058
Department of Health and Human Services	—	—	—	—	—	35	40	47	54	56	61	62	57	50	47	4	5	4	4	14	10	8	8	8	8
Department of Housing and Urban Development	—	—	—	—	—	—	10	10	—	—	—	—	—	—	—	—	—	—	—	—	0	0	<0.5	7	7
Department of Interior	22	29	27	26	26	26	27	27	27	26	29	29	29	27	27	34	45	128	168	147	55	54	58	57	63
Department of State	1	1	1	3	3	5	7	7	7	7	6	6	7	12	41	59	55	199	149	133	126	127	147	595	632
Department of Transportation	—	—	5	6	13	5	3	—	—	—	27	9	3	17	18	20	45	128	80	92	67	39	28	41	54
Department of Treasury	0	12	35	14	14	18	60	14	54	43	56	52	44	46	46	46	46	421	744	377	380	356	346	331	350
Environmental Protection Agency	26	73	124	114	99	103	126	124	146	136	124	127	130	128	121	131	139	164	172	144	137	135	145	150	170
Millennium Challenge Corporation	—	—	—	—	—	—	—	—	—	—	—	—	—	—	—	—	—	2	25	41	0	103	352	219	0
National Aeronautics and Space Administration	888	999	1,305	1,218	1,218	1,210	1,155	1,161	1,176	1,090	1,299	1,548	1,449	1,082	1,223	1,221	1,205	1,195	1,422	1,686	1,682	1,744	1,780	1,872	1,965
National Science Foundation	124	142	222	216	222	214	222	229	181	189	212	226	209	215	230	229	293	348	403	680	668	686	703	712	861
Nuclear Regulatory Commission	—	—	—	—	—	—	—	—	—	—	—	—	—	—	—	—	—	—	—	83	68	91	69	86	90
Peace Corps	—	—	—	—	—	—	—	—	—	—	—	—	—	—	—	—	—	—	—	—	12	12	12	12	12
Smithsonian Institution	7	7	7	7	7	7	7	7	7	6	6	6	6	6	6	6	6	7	11	8	8	8	8	8	9
Tennessee Valley Authority	—	—	—	—	—	—	—	—	—	—	—	—	—	—	—	—	—	—	—	9	11	10	10	10	16

Appendix IV. (Continued)

Agency	1993	1994	1995	1996	1997	1998	1999	2000	2001	2002	2003	2004	2005	2006	2007	2008	2009	2010	2011	2012	2013	2014	2015	2016	2017
U.S. Agency for International Development	200	173	192	175	147	163	236	156	157	179	214	195	183	190	114	136	222	383	491	348	334	351	331	304	352
U.S. Trade and Development Agency	—	—	—	—	—	—	16	—	—	—	—	—	—	—	—	—	10	17	21	16	19	22	27	32	42
Total	**2,352**	**2,668**	**3,271**	**2,952**	**2,876**	**3,114**	**3,535**	**3,511**	**3,603**	**3,522**	**4,584**	**5,090**	**4,900**	**4,716**	**5,499**	**6,255**	**7,757**	**8,731a**	**9,788**	**9,649**	**9,062**	**10,045**	**10,689**	**11,627**	**13,156**

Source: GAO analysis of Office of Management and Budget (OMB) reports. | GAO-18-223.

Notes: Dashes indicate that OMB did not report a value for the account for that year. Zeros indicate the funding that OMB reported for the agency for that year. Totals may not add due to rounding.

The table presents actual budget authority except for fiscal years 2011, 2013, and 2017. For fiscal years 2011 and 2017, OMB reported proposed budget authority. For fiscal year 2013, OMB reported final operating level funding. Funding is reported in nominal dollars, which are not adjusted for inflation.

[a]Total 2010 funding of $8.731 billion as presented by OMB and reported in this appendix does not match the total of $8.771 billion presented in appendixes III and V. In its June 2010 report that presented these data, OMB noted that totals may not add due to rounding and subtraction of double-counts.

Appendix V: Climate Change Funding by Line Item as Reported by the Office and Management and Budget (OMB), from Fiscal Years 1993 through 2017 (Dollars in Millions)

ACCOUNT	1993	1994	1995	1996	1997	1998	1999	2000	2001	2002	2003	2004	2005	2006	2007	2008	2009	2010	ARRAa	2011	2012	2013	2014	2015	2016	2017
TECHNOLOGY																										
Direct Technology																										
Department of Agriculture	—	—	—	—	—	—	0	0	3	3	42	45	48	49	48	51	271	453	0	265	275	164	400	319	306	319
Agricultural Research Service	—	—	—	—	—	—	0	—	0	—	2	2	2	2	2	2	5	5	0	1	—	—	—	—	—	—
Salaries and Expenses[b]	—	—	—	—	—	—	—	—	—	—	—	—	—	—	—	—	—	—	—	—	33	32	32	33	30	31
Bioenergy Research[b]	—	—	—	—	—	—	—	—	—	—	—	—	—	—	—	—	—	—	—	4	—	—	—	—	—	—
Rural Business Service																										
Renewable Energy Program	—	—	—	—	—	—	—	—	—	0	22	23	23	23	23	—	—	—	—	—	—	—	—	—	—	—
Value Added Producer Grants (Cooperative Development Grants)[c]	—	—	—	—	—	—	—	—	—	—	—	—	2	3	3	5	6	4	0	4	1	1	6	1	1	1
Rural Energy For America[c]	—	—	—	—	—	—	—	—	—	—	—	—	—	—	—	36	5	39	0	39	3	3	53	48	47	69
Biorefinery Assistance Programc	—	—	—	—	—	—	—	—	—	—	—	—	—	—	—	0	0	0	0	17	—	—	—	—	—	—
Biorefinery Program for Advanced Biofuels[c]	—	—	—	—	—	—	—	—	—	—	—	—	—	—	—	—	55	55	0	—	—	—	—	—	—	—
Forest Service																										
Forest and Rangeland Research	—	—	—	—	—	—	0	0	3	—	1	0	2	2	2	1	1	5	0	—	—	0	6	6	6	6
Research and Development —Inventories of Carbon Biomass	—	—	—	—	—	—	—	—	—	—	1	0	1	1	1	—	—	—	—	—	—	—	—	—	—	—
Biofuels/Biomass, Forest and Rangeland Research[b]	—	—	—	—	—	—	—	—	—	—	—	—	—	—	—	—	—	—	—	5	—	—	—	—	—	—
Commerciali zation/ Renewable Energy[b]	—	—	—	—	—	—	—	—	—	—	—	—	—	—	—	—	—	—	—	—	26	22	21	20	26	26

Appendix V. (Continued)

ACCOUNT	1993	1994	1995	1996	1997	1998	1999	2000	2001	2002	2003	2004	2005	2006	2007	2008	2009	2010	ARRAa	2011	2012	2013	2014	2015	2016	2017
Natural Resources Conservation Service																										
Carbon Cycle	—	—	—	—	—	—	—	—	—	—	1	1	1	1	1	0	9	7	0	7	—	—	—	—	—	—
Biomass Research and Development	—	—	—	—	—	—	—	—	—	3	14	14	13	14	14	0	—	—	—	—	—	—	—	—	—	—
Conservation Operations[b]	—	—	—	—	—	—	—	—	—	—	—	—	—	—	—	—	—	—	—	—	6	< 0.5	< 0.5	< 0.5	< 0.5	< 0.5
Farm Security and Rural Investment Programsb,[d]	—	—	—	—	—	—	—	—	—	—	—	—	—	—	—	—	—	—	—	—	16	14	13	13	13	13
Cooperative State Research, Education and Extension Service																										
Biofuels/ Biomass Research, Formula Funds, National Research Initiative	—	—	—	—	—	—	—	—	—	—	3	5	5	3	3	5	5	5	0	5	—	—	—	—	—	—
Office of the Chief Economist																										
Methane to Markets[c]	—	—	—	—	—	—	—	—	—	—	—	—	—	0	0	—	—	—	—	—	—	—	—	—	—	—
Salaries and Expenses[b]	—	—	—	—	—	—	—	—	—	—	—	—	—	—	—	—	—	—	—	—	4	3	5	2	2	2
National Agricultural Statistics Service[c]	—	—	—	—	—	—	—	—	—	—	—	—	—	—	—	—	0	0	0	—	—	—	—	—	—	—
Rural Business Service																										
Rural Energy For Americac,[d]	—	—	—	—	—	—	—	—	—	—	—	—	—	—	—	0	55	60	0	70	22	0	0	0	0	0
Repowering Assistance Programc,[d]	—	—	—	—	—	—	—	—	—	—	—	—	—	—	—	0	35	0	0	0	—	—	—	—	—	—
Biorefinery Assistance Programc,[d]	—	—	—	—	—	—	—	—	—	—	—	—	—	—	—	0	75	245	0	0	—	0	100	30	27	0
Energy Assistance Payments (formerly titled Bioenergy Program for Advanced Biofuels)[b,d]	—	—	—	—	—	—	—	—	—	—	—	—	—	—	—	—	—	—	—	85	65	0	27	14	15	15
Rural Business Cooperative Serviceb																										
Guaranteed Business and Industry Loans[b]	—	—	—	—	—	—	—	—	—	—	—	—	—	—	—	—	—	—	—	—	4	5	7	8	4	4
Rural Economic Development Loansb	—	—	—	—	—	—	—	—	—	—	—	—	—	—	—	—	—	—	—	—	0	< 0.5	0	< 0.5	< 0.5	1
National Institute of Food and Agriculture																										

ACCOUNT	1993	1994	1995	1996	1997	1998	1999	2000	2001	2002	2003	2004	2005	2006	2007	2008	2009	2010	ARRAa	2011	2012	2013	2014	2015	2016	2017
Biomass Research and Development[c],[d]	—	—	—	—	—	—	—	—	—	—	—	—	—	—	—	2	20	28	0	28	40	0	3	3	3	3
Research and Education Activities[b]	—	—	—	—	—	—	—	—	—	—	—	—	—	—	—	—	—	—	—	—	31	56	49	53	48	67
Economic Research Service[b]	—	—	—	—	—	—	—	—	—	—	—	—	—	—	—	—	—	—	—	—	2	2	2	2	2	2
Rural Utilities Service[b]																										
High Cost Energy Grants[b]	—	—	—	—	—	—	—	—	—	—	—	—	—	—	—	—	—	—	—	—	4	4	4	8	10	0
Rural Energy Savings Program Account (Rural Energy for America Sec. 9007)[b]	—	—	—	—	—	—	—	—	—	—	—	—	—	—	—	—	—	—	—	—	—	—	—	0	8	0
Farm Service Agency																										
Biomass Crop Assistance Program[b,d]	—	—	—	—	—	—	—	—	—	—	—	—	—	—	—	—	—	—	—	—	17	0	25	23	3	25
Commodity Credit Corporation[b,d]	—	—	—	—	—	—	—	—	—	—	—	—	—	—	—	—	—	—	—	—	0	22	47	56	60	54
Department of Commerce	—	—	—	—	—	—	—	—	—	—	—	—	26	18	22	14	15	18	4	18	40	48	55	55	54	55
National Institutes of Standards and Technology	—	—	—	—	—	—	—	—	—	—	40	28	26	18	—	—	—	—	—	—	—	—	—	—	—	—
Industrial Technical Services— Advanced Technology Program	—	—	—	—	—	—	—	—	—	—	30	18	8	10	6	4	7	15	4	—	—	—	—	—	—	—
Industrial Technical Services— Technology Innovation Program[b]	—	—	—	—	—	—	—	—	—	—	—	—	—	—	—	—	—	—	—	1	—	—	—	—	—	—
Scientific and Technical Services Research Services	—	—	—	—	—	—	0	2	—	—	10	10	18	8	—	—	—	—	—	15	40	48	54	54	54	54
International Trade Administration																										
Operations and Administration[c]	—	—	—	—	—	—	—	—	—	—	—	—	0	0	0	2	2	2	0	2	—	—	—	—	—	—
National Oceanic and Atmospheric Administration Operations[b]																										

Appendix V. (Continued)

ACCOUNT	1993	1994	1995	1996	1997	1998	1999	2000	2001	2002	2003	2004	2005	2006	2007	2008	2009	2010	ARRAa	2011	2012	2013	2014	2015	2016	2017
Research and Facilities[b]	—	—	—	—	—	—	—	—	—	—	—	—	—	—	—	—	—	—	—	—	0	1	1	1	0	1
Department of Defense	—	—	—	—	—	—	—	—	—	—	83	51	59	77	101	176	261	226	139	143	481	458	469	486	618	643
Research, Development, Test and Evaluation, Army	—	—	—	—	—	—	—	—	—	—	45	15	27	49	69	98	98	93	44	28	32	34	37	36	32	38
Research, Development, Test and Evaluation, Navy	—	—	—	—	—	—	—	—	—	—	16	17	18	17	13	44	54	13	18	11	231	176	210	194	124	152
Research, Development, Test and Evaluation, Air Force	—	—	—	—	—	—	—	—	—	—	3	1	1	0	13	34	108	120	35	104	118	205	181	216	427	418
Research, Development, Test and Evaluation, Defense-wide	—	—	—	—	—	—	—	—	—	—	19	19	13	11	6	0	0	0	42	0	101	43	41	40	35	36
Defense Advanced Research Projects Agency[c]	—	—	—	—	—	—	—	—	—	—	—	17	11	7	6	—	—	—	—	—	—	—	—	—	—	—
Office of the Secretary of Defensec	—	—	—	—	—	—	—	—	—	—	—	2	2	4	0	—	—	—	—	—	—	—	—	—	—	—
Department of Energy	595	753	829	683	658	729	890	980	1,050	1,519	2,099	2,390	2,342	2,374	3,032	3,663	4,543	4,399	25,223	4,883	4,388	4,195	4,714	5,196	5,606	6,816
Energy Conservation	—	—	—	—	—	—	—	—	—	897	880	868	—	—	—	—	—	—	—	—	—	—	—	—	—	—
Energy Conservation Research and Development	346	435	468	415	414	457	518	577	619	622	—	—	—	—	—	—	—	—	—	—	—	—	—	—	—	—
State Energy Grants	—	—	—	—	—	—	—	—	—	45	—	—	—	—	—	—	—	—	—	—	—	—	—	—	—	—
Weatherization	—	—	—	—	—	—	—	—	—	230	—	—	—	—	—	—	—	—	—	—	—	—	—	—	—	—
Energy Supply/Energy Supply and Conservation	249	318	361	268	244	272	332	315	375	400	667	—	—	—	—	—	—	—	—	—	—	—	—	—	—	—
Nuclear Energy Research Initiative	—	—	—	—	—	—	—	—	—	32	—	—	—	—	—	—	—	—	—	—	—	—	—	—	—	—
Electricity/Electricity Supply and Distribution/Electricity Delivery and Energy Reliability	—	—	—	—	—	—	—	—	—	—	88	73	57	77	120	130	113	109	4340	143	133	123	140	140	196	226
Renewables	249	318	361	268	244	272	332	310	370	368	322	352	—	—	—	—	—	—	—	—	—	—	—	—	—	—
Nuclear	—	—	—	—	—	—	0	5	5	—	257	309	291	343	513	682	787	747	0	793	772	738	873	884	936	912

ACCOUNT	1993	1994	1995	1996	1997	1998	1999	2000	2001	2002	2003	2004	2005	2006	2007	2008	2009	2010	ARRAa	2011	2012	2013	2014	2015	2016	2017
Energy Efficiency and Renewable Energy[c]	—	—	—	—	—	—	—	—	—	—	—	—	1,234	1,166	1,411	1,722	2,179	2,242	16,800	2,355	1,819	1,772	1,913	1,937	2,073	2,898
Fossil Energy Research and Development	—	—	—	—	—	—	24	52	18	184	253	455	374	397	493	611	762	560	3,400	—	—	—	—	—	—	—
Sequestration Research and Development	—	—	—	—	—	—	—	—	—	32	—	—	—	—	—	—	—	—	—	—	—	—	—	—	—	—
Greenhouse Gas Emission Reduction	—	—	—	—	—	—	—	—	—	152	—	—	—	—	—	—	—	—	—	—	—	—	—	—	—	—
Efficiency and Sequestration[b]	—	—	—	—	—	—	—	—	—	—	—	—	—	—	—	—	—	—	—	545	—	—	—	—	—	—
Carbon Capture and Storage and Power Systems[b]	—	—	—	—	—	—	—	—	—	—	—	—	—	—	—	—	—	—	—	—	472	436	490	503	533	564
Science	—	—	—	—	—	—	13	33	35	35	298	333	386	391	487	512	700	723	283	—	—	—	—	—	—	—
Sequestration	—	—	—	—	—	—	—	—	—	32	—	—	—	—	—	—	—	—	—	—	—	—	—	—	—	—
Fusion, Sequestration, and Hydrogen[b]	—	—	—	—	—	—	—	—	—	—	—	—	—	—	—	—	—	—	—	738	902	859	1,007	1,441	1,578	1,853
Energy Information Administration	—	—	—	—	—	—	3	3	3	3	—	—	—	—	—	—	—	—	—	—	—	—	—	—	—	—
Departmental Administration—Climate Change Technology Program Direction[c]	—	—	—	—	—	—	—	—	—	—	—	0	0	0	1	1	2	9	0	9	—	—	—	—	—	—
Innovative Technology Loan Guarantee Program[c]	—	—	—	—	—	—	—	—	—	—	—	—	—	0	7	5	—	—	—	—	—	—	—	—	—	—
Advanced Research Projects Agency	—	—	—	—	—	—	—	—	—	—	—	—	—	—	—	—	—	—	—	—	—	—	—	—	—	—
Energy (ARPA-E)c	—	—	—	—	—	—	—	—	—	—	—	—	—	—	—	—	0	9	0	300	275	251	280	281	291	350
Energy (ARPA-E) Recovery Act[c]	—	—	—	—	—	—	—	—	—	—	—	—	—	—	—	0	0	0	400	0	—	—	—	—	—	—
Bonneville Power Administration Fund[b]	—	—	—	—	—	—	—	—	—	—	—	—	—	—	—	—	—	—	—	—	15	17	10	10	0	13

Appendix V. (Continued)

ACCOUNT	1993	1994	1995	1996	1997	1998	1999	2000	2001	2002	2003	2004	2005	2006	2007	2008	2009	2010	ARRAa	2011	2012	2013	2014	2015	2016	2017
Race to the Top for Energy Efficiency and Grid Modernization[b]	—	—	—	—	—	—	—	—	—	—	—	—	—	—	—	—	—	—	—	—	0	0	0	0	0	0
HomeStar[b]	—	—	—	—	—	—	—	—	—	—	—	—	—	—	—	—	—	—	—	—	0	0	0	0	0	0
Energy Security Trust[b]	—	—	—	—	—	—	—	—	—	—	—	—	—	—	—	—	—	—	—	—	0	0	0	0	0	0
Environmental Protection Agency	—	43	102	96	86	90	109	103	123	115	102	110	110	109	105	114	111	133	0	140	117	111	108	120	122	139
Environmental Programs and Management	—	35	91	81	70	73	72	76	96	89	82	89	91	90	92	97	94	113	0	123	99	95	100	95	95	108
Science and Technology	—	8	11	15	16	17	37	27	27	26	20	22	19	19	13	17	17	20	0	17	18	16	8	9	8	11
Department of Housing and Urban Development	—	—	—	—	—	—	—	—	—	—	—	—	—	—	—	—	—	—	—	—	—	—	—	< 0.5	7	7
Research and Technology	—	—	—	—	—	—	10	10	—	—	—	—	—	—	—	—	—	—	—	—	—	—	—	—	—	—
Management and Administration--Transformation Initiative[b]	—	—	—	—	—	—	—	—	—	—	—	—	—	—	—	—	—	—	—	—	—	—	—	0	3	3
Policy Development and Research—Research and Technology[b]	—	—	—	—	—	—	—	—	—	—	—	—	—	—	—	—	—	—	—	—	—	—	—	< 0.5	7	7
Department of Interior	—	—	—	—	—	—	—	—	—	—	1	1	2	—	—	—	—	—	—	—	—	—	—	—	—	—
U.S. Geological Survey—Surveys, Investigations and Research—Geology Discipline, Energy Program	—	—	—	—	—	—	—	—	—	—	—	—	—	—	—	—	—	—	—	—	—	287	313	363	338	348
National Aeronautics and Space Administration	—	—	—	—	—	—	—	—	—	—	152	227	208	129	139	137	119	124	31	137	—	—	—	—	—	—
Exploration, Science & Aeronautics																										
Aeronautics[b]	—	—	—	—	—	—	—	—	—	—	—	—	—	—	—	—	—	—	—	—	259	255	282	323	305	321
Exploration[b]	—	—	—	—	—	—	—	—	—	—	—	—	—	—	—	—	—	—	—	—	9	6	5	5	3	4

ACCOUNT	1993	1994	1995	1996	1997	1998	1999	2000	2001	2002	2003	2004	2005	2006	2007	2008	2009	2010	ARRAa	2011	2012	2013	2014	2015	2016	2017
Space Technology[b]	—	—	—	—	—	—	—	—	—	—	—	—	—	—	—	—	—	—	—	—	28	25	26	36	29	23
National Science Foundation																										
Research and Related Activities	—	—	—	—	—	—	—	—	—	—	9	11	11	18	21	22	24	26	2	27	341	346	370	370	372	512
Department of Transportation	—	—	—	—	—	—	—	—	—	—	27	5	2	16	17	19	43	125	100	77	91	67	39	27	40	54
Federal Transit Administration																										
Capital Investment Grants	—	—	—	—	—	—	—	—	—	—	26	—	—	—	—	—	—	—	—	—	—	—	—	—	—	—
Research and University Centers and Formula and Bus Grants[c]	—	—	—	—	—	—	—	—	—	—	—	—	—	14	16	18	19	94	100	53	52	33	0	0	0	0
Office of the Secretary of Technology																										
Transportation, Policy, Research and Development	—	—	—	—	—	—	—	—	—	—	1	4	1	—	—	—	—	—	—	—	—	—	—	—	—	—
Research and Special Programs Administration																										
Research and Special Programs	—	—	—	—	—	—	—	—	—	—	0	—	—	—	—	—	—	—	—	—	—	—	—	—	—	—
National Highway Traffic Safety Administration	—	—	—	—	—	—	—	—	—	—	—	0	< 0.5	1	1	1	0	1	0	0	10	8	10	0	11	12
Research and Innovative Technology Administration																										
Research and Development	—	—	—	—	—	—	—	—	—	—	—	1	1	1	1	1	1	1	0	1	1	1	1	1	0	1
Federal Highways Administration																										
Federal-aid Highways[c]	—	—	—	—	—	—	—	—	—	—	—	—	—	—	—	—	19	19	0	19	—	—	—	—	—	—
Federal Aviation Administration																										
Research, Engineering, and Development[c]	—	—	—	—	—	—	—	—	—	—	—	—	—	—	—	—	3	11	0	5	21	20	21	22	25	25

Appendix V. (Continued)

ACCOUNT	1993	1994	1995	1996	1997	1998	1999	2000	2001	2002	2003	2004	2005	2006	2007	2008	2009	2010	ARRAa	2011	2012	2013	2014	2015	2016	2017
Facilities and Equipment[b]	—	—	—	—	—	—	—	—	—	—	—	—	—	—	—	—	—	—	—	—	7	4	6	4	0	0
Federal Railroad Association																										
Railroad Research and Development[b]	—	—	—	—	—	—	—	—	—	—	—	—	—	—	—	—	—	—	—	—	1	1	2	0	3	15
Pipelines and Hazardous Materials Safety Administration																										
Hazardous Materials Safety[b]	—	—	—	—	—	—	—	—	—	—	—	—	—	—	—	—	—	—	—	—	—	—	—	1	1	1
Nuclear Regulatory Commission																										
Salaries and Expenses[b]	—	—	—	—	—	—	—	—	—	—	—	—	—	—	—	—	—	—	—	—	83	68	91	69	86	90
Tennessee Valley Authority[b]	—	—	—	—	—	—	—	—	—	—	—	—	—	—	—	—	—	—	—	—	9	11	10	10	10	16
Direct Technology Total	595	796	931	779	744	819	1,009	1,095	1,176	1,637	2,555	2,868	2,808	2,789	3,485	4,196	5,386	5,504	25,499	5,690	6,121	5,755	6,567	6,999	7,539	8,978
National Climate Change Technology Initiative[c]																										
Department of Energy	—	—	—	—	—	—	—	—	—	—	—	—	162	289	—	—	—	—	—	—	—	—	—	—	—	—
Energy Supply and Conservation																										
Energy Efficiency and Renewable Energy[c]	—	—	—	—	—	—	—	—	—	—	—	—	65	66	—	—	—	—	—	—	—	—	—	—	—	—
Nuclear[c]	—	—	—	—	—	—	—	—	—	—	—	—	9	102	—	—	—	—	—	—	—	—	—	—	—	—
Fossil Energy Research and Development																										
Efficiency and Sequestration[c]	—	—	—	—	—	—	—	—	—	—	—	—	89	121	—	—	—	—	—	—	—	—	—	—	—	—
Departmental Administration																										
Climate Change Technology Program Direction[c]	—	—	—	—	—	—	—	—	—	—	—	—	0	0	—	—	—	—	—	—	—	—	—	—	—	—
Environmental Protection Agency																										
Environmental Programs and Management[c]	—	—	—	—	—	—	—	—	—	—	—	—	11	10	—	—	—	—	—	—	—	—	—	—	—	—
National Climate Change Technology Initiative Total	—	—	—	—	—	—	—	—	—	—	—	—	173	299	—	—	—	—	—	—	—	—	—	—	—	—

ACCOUNT	1993	1994	1995	1996	1997	1998	1999	2000	2001	2002	2003	2004	2005	2006	2007	2008	2009	2010	ARRAa	2011	2012	2013	2014	2015	2016	2017
Indirect Technology																										
Department of Energy	250	242	231	212	201	351	417	434	499	—	—	—	—	—	—	—	—	—	—	—	—	—	—	—	—	—
Fossil Energy Research and Development	250	242	231	212	201	196	233	242	274	—	—	—	—	—	—	—	—	—	—	—	—	—	—	—	—	—
Coal-Efficient Combustion & Utilization	186	166	144	120	101	105	—	—	—	—	—	—	—	—	—	—	—	—	—	—	—	—	—	—	—	—
Natural Gas-- Efficient Combustion & Utilization	64	76	87	92	100	91	—	—	—	—	—	—	—	—	—	—	—	—	—	—	—	—	—	—	—	—
Energy Supply																										
Nuclear Energy Research and Development	—	—	—	—	—	0	18	22	34	—	—	—	—	—	—	—	—	—	—	—	—	—	—	—	—	—
Energy Conservation Research and Development																										
Weatherization & State Energy Grants	—	—	—	—	—	155	166	169	191	—	—	—	—	—	—	—	—	—	—	—	—	—	—	—	—	—
Biobased Products & Bioenergy	—	—	—	—	—	—	195	200	—	—	—	—	—	—	—	—	—	—	—	—	—	—	—	—	—	—
Department of Agriculture	—	—	—	—	—	—	86	76	—	—	—	—	—	—	—	—	—	—	—	—	—	—	—	—	—	—
Agriculture Research Service	—	—	—	—	—	—	44	46	—	—	—	—	—	—	—	—	—	—	—	—	—	—	—	—	—	—
Cooperative State Research, Education, & Extension Service																										
Research and Education Assistance	—	—	—	—	—	—	11	11	—	—	—	—	—	—	—	—	—	—	—	—	—	—	—	—	—	—
Initiative for Future Agriculture & Food Systems	—	—	—	—	—	—	—	9	—	—	—	—	—	—	—	—	—	—	—	—	—	—	—	—	—	—
Forest Service																										
Forest and Rangeland Management	—	—	—	—	—	—	9	9	—	—	—	—	—	—	—	—	—	—	—	—	—	—	—	—	—	—

Appendix V. (Continued)

ACCOUNT	1993	1994	1995	1996	1997	1998	1999	2000	2001	2002	2003	2004	2005	2006	2007	2008	2009	2010	ARRAa	2011	2012	2013	2014	2015	2016	2017
Executive Operations	—	—	—	—	—	—	1	1	—	—	—	—	—	—	—	—	—	—	—	—	—	—	—	—	—	—
Departmental Administration	—	—	—	—	—	—	< 0.5	< 0.5	—	—	—	—	—	—	—	—	—	—	—	—	—	—	—	—	—	—
Alternative Agricultural Research and Commercialization	—	—	—	—	—	—	4	—	—	—	—	—	—	—	—	—	—	—	—	—	—	—	—	—	—	—
Natural Resources Conservation Service																										
Forestry Incentives Program	—	—	—	—	—	—	16	—	—	—	—	—	—	—	—	—	—	—	—	—	—	—	—	—	—	—
Rural Development																										
Rural Community Advancement Program	—	—	—	—	—	—	1	—	—	—	—	—	—	—	—	—	—	—	—	—	—	—	—	—	—	—
Department of Energy	—	—	—	—	—	—	109	124	—	—	—	—	—	—	—	—	—	—	—	—	—	—	—	—	—	—
Energy Supply																										
Solar and Renewable Energy Research and Development	—	—	—	—	—	—	40	70	—	—	—	—	—	—	—	—	—	—	—	—	—	—	—	—	—	—
Energy Conservation Research and Development	—	—	—	—	—	—	41	11	—	—	—	—	—	—	—	—	—	—	—	—	—	—	—	—	—	—
Fossil Energy Research and Development	—	—	—	—	—	—	0	13	—	—	—	—	—	—	—	—	—	—	—	—	—	—	—	—	—	—
Science (Basic Science)	—	—	—	—	—	—	27	30	—	—	—	—	—	—	—	—	—	—	—	—	—	—	—	—	—	—
Partnership for a New Generation of Vehicles	—	—	—	—	—	—	73	64	—	—	—	—	—	—	—	—	—	—	—	—	—	—	—	—	—	—
Department of Commerce	—	—	63	56	42	29	30	22	—	—	—	—	—	—	—	—	—	—	—	—	—	—	—	—	—	—
Under Secretary for Technology/ Office of Technology Policy	—	—	—	—	—	—	1	0	—	—	—	—	—	—	—	—	—	—	—	—	—	—	—	—	—	—
Salaries and Expenses	—	—	0	1	1	1	—	—	—	—	—	—	—	—	—	—	—	—	—	—	—	—	—	—	—	—
National Institutes of Standards and Technology	—	—	—	—	—	—	29	22	—	—	—	—	—	—	—	—	—	—	—	—	—	—	—	—	—	—
Scientific and Technical Research and Services	—	—	—	7	7	7	6	—	—	—	—	—	—	—	—	—	—	—	—	—	—	—	—	—	—	—

ACCOUNT	1993	1994	1995	1996	1997	1998	1999	2000	2001	2002	2003	2004	2005	2006	2007	2008	2009	2010	ARRAa	2011	2012	2013	2014	2015	2016	2017
Industrial Technology Services	—	—	—	56	48	34	22	—	—	—	—	—	—	—	—	—	—	—	—	—	—	—	—	—	—	—
National Science Foundation																										
Research and Related Activities	—	—	53	53	56	47	40	42	—	—	—	—	—	—	—	—	—	—	—	—	—	—	—	—	—	—
Department of Transportation																										
National Highway Traffic Safety Administration (and FTA prior to FY 1999)																										
Operations and Research	—	—	5	6	13	5	3	—	—	—	—	—	—	—	—	—	—	—	—	—	—	—	—	—	—	—
Indirect Technology Total	250	242	352	327	312	432	685	698	499	e	e	e	e	e	e	e	e	e	e	e	e	e	e	e	e	e
Technology Total	845	1,038	1,283	1,106	1,056	1,251	1,694	1,793	1,675	1,637	2,555	2,868	2,808	2,789	3,485	4,196	5,386	5,504	25,499	5,690	6,121	5,755	6,567	6,999	7,539	8,978
U.S. Global Change Research Program																										
SCIENCE																										
Department of Agriculture	55	56	60	52	57	53	52	56	51	56	60	70	62	61	61	65	47	109	0	157	116	108	115	97	98	124
Agricultural Research Service	17	18	24	24	26	27	26	28	29	30	35	37	38	38	40	39	20	20	0	22	36	38	38	45	45	65
Cooperative State Research, Education, & Extension Services																										
Research and Education	11	12	10	10	12	7	7	9	4	9	8	16	5	4	2	4	—	—	—	—	—	—	—	—	—	—
Economic Research Service	1	1	1	1	1	1	1	1	1	<0.5	0	0	0	0	0	0	0	1	0	1	2	2	2	1	2	3
Natural Resources Conservation Services	—	—	—	—	—	—	—	—	—	—	—	—	—	—	—	—	—	—	—	—	1	1	1	2	3	3
Conservation Operations	2	2	2	2	1	1	1	1	—	—	—	—	—	—	—	—	—	—	—	—	—	—	—	—	—	—
Forest Service																										
Forest and Rangeland Research	24	23	23	15	17	17	17	17	17	17	17	17	18	18	19	22	22	32	0	28	26	26	27	25	28	28

Appendix V. (Continued)

ACCOUNT	1993	1994	1995	1996	1997	1998	1999	2000	2001	2002	2003	2004	2005	2006	2007	2008	2009	2010	ARRAa	2011	2012	2013	2014	2015	2016	2017
National Institute of Food and Agriculture[c]	—	—	—	—	—	—	—	—	—	—	—	—	—	—	—	—	5	56	0	106	50	40	43	20	15	19
National Agricultural Statistics Service[b]	—	—	—	—	—	—	—	—	—	—	—	—	—	—	—	—	—	—	—	—	1	1	1	1	1	1
Animal and Plant Health Inspection Service[b]	—	—	—	—	—	—	—	—	—	—	—	—	—	—	—	—	—	—	—	—	—	—	—	0	< 0.5	< 0.5
Farm Service Agency[b]	—	—	—	—	—	—	—	—	—	—	—	—	—	—	—	—	—	—	—	—	—	—	—	0	1	1
Office of Chief Economist[b]	—	—	—	—	—	—	—	—	—	—	—	—	—	—	—	—	—	—	—	—	—	—	—	3	3	3
Department of Commerce	—	—	—	—	—	—	—	—	—	—	—	—	—	235	236	272	377	360	218	437	319	301	329	312	283	342
National Oceanic and Atmospheric Administration																										
Operations, Research, and Facilities	66	63	57	57	60	60	63	67	93	100	98	116	120	226	229	265	274	309	0	315	245	233	285	264	252	307
Procurement, Acquisition and Construction[c]	—	—	—	—	—	—	—	—	—	—	—	—	—	9	7	7	101	49	218	120	69	64	40	44	28	31
National Institute of Standards and Technology	—	—	—	—	—	—	—	—	—	—	—	—	—	—	0	0	2	2	0	2	5	4	4	4	4	4
Department of Energy																										
Science—Biological & Environmental Research	118	118	113	113	109	106	114	114	116	117	112	129	127	130	126	128	168	165	65	191	211	209	217	214	238	242
Environmental Protection Agency																										
Science and Technology	26	30	22	18	13	13	17	21	23	21	22	17	20	19	16	17	18	21	0	22	18	17	18	16	19	22
Department of Health and Human Services	—	—	—	—	—	35	—	—	—	—	—	—	—	—	—	—	—	—	—	—	14	10	8	8	8	8
National Institutes of Health	—	—	—	—	—	35	40	47	54	56	61	62	57	50	47	4	5	4	0	4	8	10	8	8	8	8
National Institute of Environmental Health Sciences	—	—	—	—	—	4	—	—	—	—	—	—	—	—	—	—	—	—	—	—	—	—	—	—	—	—
National Eye Institutes	—	—	—	—	—	9	—	—	—	—	—	—	—	—	—	—	—	—	—	—	—	—	—	—	—	—
National Cancer Institute	—	—	—	—	—	21	—	—	—	—	—	—	—	—	—	—	—	—	—	—	—	—	—	—	—	—
National Institute of Arthritis and Musculoskeletal & Skin Diseases	—	—	—	—	—	< 0.5	—	—	—	—	—	—	—	—	—	—	—	—	—	—	—	—	—	—	—	—

ACCOUNT	1993	1994	1995	1996	1997	1998	1999	2000	2001	2002	2003	2004	2005	2006	2007	2008	2009	2010	ARRAa	2011	2012	2013	2014	2015	2016	2017
Centers for Disease Control and Prevention	—	—	—	—	—	—	—	—	—	—	—	—	—	—	—	—	—	—	—	—	6	—	—	—	—	—
Department of the Interior																										
U.S. Geological Survey																										
Surveys, Investigations, and Research	22	29	27	26	26	26	27	27	27	26	28	28	27	27	27	34	45	63	0	81	59	55	54	58	57	63
National Aeronautics and Space Administration																										
Science, Aeronautics, and Technology	888	999	1,305	1,218	1,218	1,210	1,155	1,161	1,176	1,090	1,144	1,321	1,241	953	1,084	1,084	—	—	—	—	—	—	—	—	—	—
Science[c]	—	—	—	—	—	—	—	—	—	—	—	—	—	—	0	0	1,086	1,071	237	1,285	1,390	1,395	1,431	1,417	1,534	1,617
National Science Foundation																										
Research and Related Activities	124	142	169	163	166	167	182	187	181	189	188	215	198	197	207	207	269	319	121	370	333	316	313	331	339	348
Smithsonian Institution																										
Salaries and Expenses	7	7	7	7	7	7	7	7	7	6	6	6	6	6	6	6	6	7	0	11	8	8	8	8	8	9

Appendix V. (Continued)

ACCOUNT	1993	1994	1995	1996	1997	1998	1999	2000	2001	2002	2003	2004	2005	2006	2007	2008	2009	2010	ARRAa	2011	2012	2013	2014	2015	2016	2017
Department of State																										
International Organizations and Programs[c]	—	—	—	—	—	—	—	—	—	—	—	1	—	—	—	—	—	—	—	39	—	—	—	—	—	—
Other- non-add[b]	—	—	—	—	—	—	—	—	—	—	—	—	—	—	—	—	—	—	—	—	3	3	3	—	—	—
Department of Transportation	—	—	—	—	—	—	—	—	—	—	—	4	1	1	1	1	2	3	0	3	1	1	1	1	1	1
Federal Highways Administration																										
Federal-aid Highways[c]	—	—	—	—	—	—	—	—	—	—	—	4	1	< 0.5	< 0.5	1	< 0.5	0	0	0	0	0	0	0	0	0
Federal Aviation Administration																										
Research, Engineering, and Development[c]	—	—	—	—	—	—	—	—	—	—	—	0	< 0.5	< 0.5	< 0.5	< 0.5	2	3	0	3	1	1	1	1	1	1
Federal Transit Administration																										
Research and University Research Centers and Formula Bus Grants[c]	—	—	—	—	—	—	—	—	—	—	—	—	—	< 0.5	< 0.5	< 0.5	< 0.5	< 0.5	0	0	0	0	0	0	0	0
U.S. Agency for International Development																										

ACCOUNT	1993	1994	1995	1996	1997	1998	1999	2000	2001	2002	2003	2004	2005	2006	2007	2008	2009	2010	ARRAa	2011	2012	2013	2014	2015	2016	2017
Development Assistance	—	—	—	—	—	—	—	—	—	6	6	6	6	13	14	14	17	36	0	43	—	—	—	—	—	—
Development Assistance- non-add[b]	—	—	—	—	—	—	—	—	—	—	—	—	—	—	—	—	—	—	—	—	11	11	11	6	9	10
U.S. Global Change Research Program Total	1,306	1,444	1,760	1,654	1,656	1,677	1,657	1,687	1,728	1,667	1,725	1,803	1,660	1,488	1,825	1,832	2,023	2,122	641	2,561	2,506	2,417	2,492	2,461	2,584	2,775
Climate Change Research Initiative																										
Department of Agriculture	—	—	—	—	—	—	—	—	—	—	2	6	8	8	—	—	—	—	—	—	—	—	—	—	—	—
Agricultural Research Service	—	—	—	—	—	—	—	—	—	—	0	1	2	2	—	—	—	—	—	—	—	—	—	—	—	—
Forest Service																										
Forest and Rangeland Research	—	—	—	—	—	—	—	—	—	—	1	5	6	6	—	—	—	—	—	—	—	—	—	—	—	—
Department of Commerce																										
National Oceanic and Atmospheric Administration																										

Appendix V. (Continued)

ACCOUNT	1993	1994	1995	1996	1997	1998	1999	2000	2001	2002	2003	2004	2005	2006	2007	2008	2009	2010	ARRAa	2011	2012	2013	2014	2015	2016	2017
Operations, Research, and Facilities	—	—	—	—	—	—	—	—	—	—	18	34	46	34	—	—	—	—	—	—	—	—	—	—	—	—
Department of Energy																										
Science—Biological & Environmental Research	—	—	—	—	—	—	—	—	—	—	3	27	25	28	—	—	—	—	—	—	—	—	—	—	—	—
National Aeronautics and Space Administration																										
Science, Aeronautics, and Technology	—	—	—	—	—	—	—	—	—	—	3	65	94	95	—	—	—	—	—	—	—	—	—	—	—	—
National Science Foundation																										
Research and Related Activities	—	—	—	—	—	—	—	—	—	—	15	30	25	25	—	—	—	—	—	—	—	—	—	—	—	—
Department of State																										
International Organizations and Programs	—	—	—	—	—	—	—	—	—	—	—	1	—	—	—	—	—	—	—	—	—	—	—	—	—	—
Department of Transportation	—	—	—	—	—	—	—	—	—	—	—	4	1	—	—	—	—	—	—	—	—	—	—	—	—	—
Federal Highway Administration																										
Federal-aid-- Highways	—	—	—	—	—	—	—	—	—	—	—	4	1	0	—	—	—	—	—	—	—	—	—	—	—	—
Federal Transit Administration																										
Formula Grants and Research	—	—	—	—	—	—	—	—	—	—	—	—	—	—	—	—	—	—	—	—	—	—	—	—	—	—
Federal Aviation Administration																										
Research, Engineering, and Development	—	—	—	—	—	—	—	—	—	—	—	0	0	—	—	—	—	—	—	—	—	—	—	—	—	—
U.S. Agency for International Development																										
Development Assistance	—	—	—	—	—	—	—	—	—	—	—	6	6	13	—	—	—	—	—	—	—	—	—	—	—	—
Climate Change Research Initiative Total	—	—	—	—	—	—	—	—	—	—	41	173	204	203	—	—	—	—	—	—	—	—	—	—	—	—
Science Total	1,306	1,444	1,760	1,654	1,656	1,677	1,657	1,687	1,728	1,667	1,766	1,976	1,864	1,691	1,825	1,832	2,023	2,122	641	2,561	2,506	2,417	2,492	2,461	2,584	2,775

ACCOUNT	1993	1994	1995	1996	1997	1998	1999	2000	2001	2002	2003	2004	2005	2006	2007	2008	2009	2010	ARRAa	2011	2012	2013	2014	2015	2016	2017
INTERNATIONAL ASSISTANCE																										
Core Agencies																										
Department of State	—	—	—	—	—	—	—	—	—	—	—	5	7	12	41	59	55	199	—	149	133	126	127	147	595	632
International Organizations and Programs	1	1	1	3	3	5	7	7	7	7	6	5	6	6	6	29	29	39	—	39	37	35	36	36	36	46
Economic Support Fund[c]	—	—	—	—	—	—	—	—	—	—	—	0	1	6	32	26	24	158	—	110	96	91	92	112	559	586
Diplomatic and Consular Affairs[c]	—	—	—	—	—	—	—	—	—	—	—	—	0	0	3	4	2	2	—	0	0	—	—	—	—	—
Department of Treasury	—	—	—	—	—	—	—	—	—	43	56	52	44	46	46	46	46	421	—	744	377	380	356	346	331	350
International Development Assistance																										
Global Environment Facility[f]	—	12	35	14	14	8	60	14	41	38	56	32	20	26	26	26	26	26	—	89	60	62	72	82	101	88
Debt Restructuring																										
Tropical Forest Conservation	—	—	—	—	—	—	—	—	13	5	—	20	24	20	20	20	20	20	—	20	12	11	0	0	0	0
Asian Development Bank[c]	—	—	—	—	—	—	—	—	—	—	—	—	0	—	—	—	—	—	—	—	—	—	—	—	—	—
Clean Technology Fund[c]	—	—	—	—	—	—	—	—	—	—	—	—	—	—	0	0	0	300	—	400	230	196	210	201	171	0
Strategic Climate Fund[c]	—	—	—	—	—	—	—	—	—	—	—	—	—	—	—	0	0	75	—	235	75	110	75	63	60	0
Green Climate Fund[b]	—	—	—	—	—	—	—	—	—	—	—	—	—	—	—	—	—	—	—	—	—	0	0	0	0	250
Central American and Caribbean Catastrophe Risk Insurance Programb	—	—	—	—	—	—	—	—	—	—	—	—	—	—	—	—	—	—	—	—	—	0	0	0	0	13
U.S. Agency for International Development	—	—	—	—	—	—	236	156	157	174	208	195	183	190	100	122	222	383	—	491	348	334	351	331	304	352
Development Assistance	200	173	192	175	147	163	169	109	112	116	140	125	134	118	89	88	113	313	—	446	322	308	296	272	268	310

Appendix V. (Continued)

ACCOUNT	1993	1994	1995	1996	1997	1998	1999	2000	2001	2002	2003	2004	2005	2006	2007	2008	2009	2010	ARRAa	2011	2012	2013	2014	2015	2016	2017
Development Credit Authority	—	—	—	—	—	—	1	1	1	—	—	—	—	—	—	—	—	—	—	—	—	—	—	—	—	—
Economic Support Fund	—	—	—	—	—	—	19	8	—	12	6	9	5	33	0	6	94	44	—	19	12	27	55	59	21	42
Assistance for the Independent States of he Former Soviet Union	—	—	—	—	—	—	35	34	31	30	48	47	34	30	5	15	—	—	—	—	—	—	—	—	—	—
Assistance for Eastern Europe and the Baltic States	—	—	—	—	—	—	12	4	13	11	8	7	5	6	3	11	—	—	—	—	—	—	—	—	—	—
Assistance for Europe, Eurasia, and Central Asia[c]	—	—	—	—	—	—	—	—	—	—	—	—	—	—	—	0	15	26	—	26	15	0	0	0	16	0
International Disaster Assistance	—	—	—	—	—	—	—	—	—	4	4	2	2	2	2	2	0	0	—	0	0	—	—	—	—	—
Andean Counterdrug Initiative	—	—	—	—	—	—	—	—	—	—	2	3	2	0	0	—	—	—	—	—	—	—	—	—	—	—
P.L. 480 Title II Food Aid[c]	—	—	—	—	—	—	—	—	—	—	—	1	1	1	0	—	—	—	—	—	—	—	—	—	—	—
Core Agencies Total	—	—	—	—	—	—	—	—	—	—	—	—	—	—	—	—	323	1,003	—	1,384	858	840	834	824	1,230	1,334
Complementary Agencies[c]																										
Department of Agriculture																										
Forest Service																										
Forest and Rangeland Research[c]	—	—	—	—	—	—	—	—	—	—	—	—	—	—	—	—	4	5	—	4	3	3	3	3	1	3
Department of Commerce	—	—	—	—	—	—	—	—	—	—	—	—	—	—	—	—	11	11	—	11	—	—	—	—	—	—
National Oceanic and Atmospheric Administration																										
Operations, Research and Facilities[c]	—	—	—	—	—	—	—	—	—	—	—	—	—	—	—	—	9	9	—	9	—	—	—	—	—	—
International Trade Administration																										
Operations and Administration[c]	—	—	—	—	—	—	—	—	—	—	—	—	—	—	—	—	2	2	—	2	—	—	—	—	—	—
Department of Energy	—	—	—	—	—	—	—	—	—	—	—	—	—	—	—	—	0	13	—	13	13					
Energy Supply																										
Solar and Renewable Energy Research and Development	—	—	—	—	—	—	6	—	—	—	—	—	—	—	—	—	—	—	—	—	—	—	—	—	—	—
Energy Efficiency and Renewable Energyc	—	—	—	—	—	—	—	—	—	—	—	—	—	—	—	—	0	8	—	8	9	—	—	—	—	—

ACCOUNT	1993	1994	1995	1996	1997	1998	1999	2000	2001	2002	2003	2004	2005	2006	2007	2008	2009	2010	ARRAa	2011	2012	2013	2014	2015	2016	2017
Fossil Energy Research and Development																										
Efficiency and Sequestration[c]	—	—	—	—	—	—	—	—	—	—	—	—	—	—	—	—	0	3	—	3	—	—	—	—	—	—
Carbon Capture and Storage and Power Systems[b]	—	—	—	—	—	—	—	—	—	—	—	—	—	—	—	—	—	—	—	—	3	—	—	—	—	—
Science[c]	—	—	—	—	—	—	—	—	—	—	—	—	—	—	—	—	0	3	—	3	1	—	—	—	—	—
Energy Programs[b]	—	—	—	—	—	—	—	—	—	—	—	—	—	—	—	—	—	—	—	—	—	13	13	9	9	9
Adjustments for programs included in multiple categories	—	—	—	—	—	—	—	—	—	—	—	—	—	—	—	—	—	—	—	—	(13)	(13)	(13)	(9)	(9)	(9)
Environmental Protection Agency																										
Environmental Programs and Management[c]	—	—	—	—	—	—	—	—	—	—	—	—	—	—	—	—	20	21	—	21	18	16	17	17	17	19
Adjustments for programs included in multiple categories	—	—	—	—	—	—	—	—	—	—	—	—	—	—	—	—	—	—	—	—	(9)	(7)	(8)	(8)	(8)	(10)
Millennium Challenge Corporationc	—	—	—	—	—	—	—	—	—	—	—	—	—	—	—	—	0	2	—	25	41	0	103	352	219	0
National Aeronautics and Space Administration																										
Science[c]	—	—	—	—	—	—	—	—	—	—	—	—	—	—	—	—	2	2	—	6	3	6	6	6	6	7
Adjustments for programs included in multiple categories	—	—	—	—	—	—	—	—	—	—	—	—	—	—	—	—	—	—	—	—	(3)	(6)	(6)	(6)	(6)	(7)
National Science Foundation																										
Research and Related Activities	—	—	—	—	—	—	—	—	—	—	—	—	—	—	—	—	3	6	—	9	6	6	4	2	1	1
Peace Corps																										
Peace Corps	—	—	—	—	—	—	—	—	—	—	—	—	—	—	—	—	—	—	—	—	—	13	12	12	12	12
U.S. Trade and Development Agency	—	—	—	—	—	—	16	—	—	—	—	—	—	—	—	—	10	17	—	21	16	19	22	27	32	42
Complementary Agencies Total	—	—	—	—	—	—	—	—	—	—	—	—	—	—	—	—	50	77	—	110	100	76	180	428	297	93
International Assistance Total	201	186	228	192	164	186	325	177	218	224	270	252	234	249	188	227	373	1,080	—	1,494	958	916	1,013	1,252	1,527	1,426
WILDLIFE ADAPTATIONg																										
Department of Interior																										
National Park Service																										

Appendix V. (Continued)

ACCOUNT	1993	1994	1995	1996	1997	1998	1999	2000	2001	2002	2003	2004	2005	2006	2007	2008	2009	2010	ARRAa	2011	2012	2013	2014	2015	2016	2017
Operation of the National Park Service[c]	—	—	—	—	—	—	—	—	—	—	—	—	—	—	—	—	0	10	0	10	—	—	—	—	—	—
Fish and Wildlife Service																										
Resource Management[c]	—	—	—	—	—	—	—	—	—	—	—	—	—	—	—	—	0	40	0	59	—	—	—	—	—	—
State and Tribal Wildlife Grants[b]	—	—	—	—	—	—	—	—	—	—	—	—	—	—	—	—	—	—	—	0	—	—	—	—	—	—
Bureau of Land Management																										
Management of Lands and Resources[c]	—	—	—	—	—	—	—	—	—	—	—	—	—	—	—	—	0	15	0	18	—	—	—	—	—	—
Bureau of Indian Affairs																										
Operation of Indian Programs[b]	—	—	—	—	—	—	—	—	—	—	—	—	—	—	—	—	—	—	—	0	—	—	—	—	—	—
Wildlife Adaptation Total	—	—	—	—	—	—	—	—	—	—	—	—	—	—	—	—	0	65	0	87	—	—	—	—	—	—
NATURAL RESOURCES ADAPTATION[e]																										
National Park Service																										
Operation of the National Park Service[b]	—	—	—	—	—	—	—	—	—	—	—	—	—	—	—	—	—	—	—	—	3	—	—	—	—	—
Fish and Wildlife Service																										
Resource Management[b]	—	—	—	—	—	—	—	—	—	—	—	—	—	—	—	—	—	—	—	—	60	—	—	—	—	—
Bureau of Land Management																										
Management of Lands and Resources[c]	—	—	—	—	—	—	—	—	—	—	—	—	—	—	—	—	—	—	—	—	18	—	—	—	—	—
Bureau of Indian Affairs																										
Operation of Indian Programs[b]	—	—	—	—	—	—	—	—	—	—	—	—	—	—	—	—	—	—	—	—	0	—	—	—	—	—
Bureau of Reclamation																										

ACCOUNT	1993	1994	1995	1996	1997	1998	1999	2000	2001	2002	2003	2004	2005	2006	2007	2008	2009	2010	ARRAa	2011	2012	2013	2014	2015	2016	2017
Cooperative Landscape Conservationb	—	—	—	—	—	—	—	—	—	—	—	—	—	—	—	—	—	—	—	—	7	—	—	—	—	—
Natural Resources Adaptation Total	—	—	—	—	—	—	—	—	—	—	—	—	—	—	—	—	—	—	—	—	88	—	—	—	—	—
Total Climate Change Funding	**2,352**	**2,668**	**3,271**	**2,952**	**2,876**	**3,114**	**3,535**	**3,511**	**3,603**	**3,522**	**4,584**	**5,090**	**5,269**	**5,876**	**5,498**	**6,255**	**7,782**	**8,771**	**26,140**	**9,788**	**9,649**	**9,062**	**10,045**	**10,689**	**11,627**	**13,156**

Source: GAO analysis of Office of Management and Budget reports. | GAO-18-223.

Notes: Dashes indicate that OMB did not report a value for the account for that year. Zeros indicate the funding that OMB reported for the account for that year. Totals may not add due to rounding.

The table presents actual budget authority except for fiscal years 2011, 2013, and 2017. For fiscal years 2011 and 2017, OMB reported proposed budget authority. For fiscal year 2013, OMB reported final operating level funding. Funding is reported in nominal dollars, which are not adjusted for inflation.

Climate-related tax expenditures are not included in this table to maintain comparability with our 2005 report, *Climate Change: Federal Reports on Climate Change Funding Should Be Clearer and More Complete*, GAO-05-461 and our 2011 report, *Climate Change: Improvements Needed to Clarify National Priorities and Better Align Them with Federal Funding Decisions*, GAO-11-317.

[a]In its June 2010 report, OMB reported funding for climate change programs and activities in the American Recovery and Reinvestment Act of 2009 (Pub. L. No. 111-5 (Feb. 17, 2009)).

[b]Funding for this account has been added since our 2011 report, *Climate Change: Improvements Needed to Clarify National Priorities and Better Align Them with Federal Funding decisions*, GAO-11-317.

[c]Funding for this account has been added since our 2005 report, *Climate Change: Federal Reports on Climate Change Funding Should Be Clearer and more Complete*, GAO-05-461.

[d]OMB identified funding for this account as mandatory under the Food, Conservation, and Energy Act of 2008 (Pub. L. No. 110-234 (May 22, 2008)).

[e]OMB did not distinguish between indirect and direct technology funding for this year.

Appendix VI: Climate Change Tax Expenditures as Reported by the Office of Management and Budget (OMB), from Fiscal Years 2003 through 2017 (Dollars in Millions)

	2003	2004	2005	2006	2007	2008	2009	2010	2011	2012	2013	2014	2015	2016	2017
Tax provisions that may reduce greenhouse gases															
Energy production credit (without coal)[a]	380	330	219	440	410	900	360	780	1,020	1,452	1,719	1,759	1,540	1,940	2,240
Credit for alternative motor vehicles and refueling property[b]	90	70	70	110	260	170	130	240	260	100	180	260	540	550	670
Exclusion of utility conservation subsidies	110	100	80	110	120	120	140	140	130	270	250	250	430	450	470
Credit for holding clean renewable energy bonds	—	—	0	20	20	40	70	80	100	70	70	70	70	70	70
Allowance of deduction for certain energy efficient commercial building property	—	—	0	80	190	170	60	80	90	70	70	40	30	(10)	(30)
Credit for construction of new energy efficient homes	—	—	0	10	20	30	30	20	20	70	40	20	760	20	0
Credit for energy efficiency improvements to existing homes	—	—	0	230	380	230	570	1,950	1,460	780	0	0	270	0	0
Credit for energy efficient appliances	—	—	0	120	80	120	130	130	50	210	300	130	—	—	—
Credit for residential energy efficient property[c]	—	—	0	10	10	20	110	180	180	910	1,010	1,140	850	770	460
Credit for business installation of qualified fuel cells	—	—	0	30	30	0	—	—	—	—	—	—	—	—	—
Energy investment credit[d]	—	—	—	—	—	40	270	530	600	1,040	1,270	1,360	1,010	1,470	970
Qualified energy conservation bonds	—	—	—	—	—	0	0	10	40	20	30	30	30	30	30
Industrial CO2 capture and sequestration tax credit	—	—	—	—	—	—	—	—	—	60	60	70	80	110	150
Advanced nuclear power production credit	—	—	—	—	—	—	—	—	—	—	—	—	0	140	140
Total energy tax provisions that may reduce greenhouse gases	580	500	369	1,160	1,520	1,840	1,870	4,140	3,950	5,052	4,999	5,129	5,610	5,540	5,170

	2003	2004	2005	2006	2007	2008	2009	2010	2011	2012	2013	2014	2015	2016	2017
Energy grants															
Energy grants in lieu of energy investment creditd,e	—	—	—	—	—	—	1,050	3,090	4,460	5,080	8,080	4,710	2,300	1,200	650
Total (tax provisions plus grants)	**580**	**500**	**369**	**1,160**	**1,520**	**1,840**	**2,920**	**7,230**	**8,410**	**10,132**	**13,079**	**9,839**	**7,910**	**6,740**	**5,820**

Source: GAO analysis of Office of Management and Budget reports. | GAO-18-223.

Notes: Dashes indicate that OMB did not report a value for the account for that year. Zeros indicate the funding that OMB reported for the agency for that year.

OMB did not report revenue effects for existing tax expenditures that may reduce greenhouse gases from fiscal years1993 through 2002. OMB began reporting revenue effects for existing climate-related tax expenditures in response to recommendations from our 2005 report, *Climate Change: Federal Reports on Climate Change Funding Should Be Clearer and more Complete*, GAO-05-461. Data for 2003 and 2004 were presented in OMB's April 2006 report. Data for 2011 was presented in OMB's June 2010 report. (Continued)

[a]Estimates of revenue loss from coal provisions have been removed from the tax expenditure estimate in the budget. In previous years, the Energy Investment Credit was contained within the New Technology Credit. Prior to OMB's August 2013 report, the energy production credit was referred to as the new technology credit (without coal).

[b]Prior to OMB's August 2013 report, the tax credit for alternative motor vehicles and refueling property was referred to as the tax credit and deduction for clean-burning vehicles.

[c]Prior to OMB's August 2013 report, the credit for residential energy efficient property was referred to as the credit for residential purchases/ installations of solar and fuel cells.

[d]In previous years, the energy investment credit was contained within the New Technology Credit. The Energy Investment Credit also includes the business installation of fuel cells, which was an independent entry in tables from previous years. Estimates of revenue loss from the micro-turbine provision have been removed from the tax expenditure estimate in the budget.

[e]Firms can take an energy grant in lieu of the energy production credit or the energy investment credit for facilities placed in service in 2009, 2010, or 2011 or whose construction commenced in 2009, 2010, or 2011. The grants are considered outlays and are direct substitutes for the energy tax provisions.

In: Climate Change
Editor: Jeremy Nesbitt
ISBN: 978-1-53617-216-4

Chapter 4

CLIMATE CHANGE: OPPORTUNITIES TO REDUCE FEDERAL FISCAL EXPOSURE*

United States Government Accountability Office

WHY GAO DID THIS STUDY

Since 2005, federal funding for disaster assistance is at least $450 billion, including approximately $19.1 billion in supplemental appropriations signed into law on June 6, 2019. In 2018 alone, there were 14 separate billion-dollar weather and climate disaster events across the United States, with a total cost of at least $91 billion, according to the National Oceanic and Atmospheric Administration. The U.S. Global Change Research Program projects that disaster costs will likely increase as certain extreme weather events become more frequent and intense due to climate change.

---

* This is an edited, reformatted and augmented version of United States Government Accountability Office; Report to Congressional Requesters, Publication No. GAO-19-625T, dated June 2019.

The costs of recent weather disasters have illustrated the need for planning for climate change risks and investing in resilience. Resilience is the ability to prepare and plan for, absorb, recover from, and more successfully adapt to adverse events, according to the National Academies of Science, Engineering, and Medicine. Investing in resilience can reduce the need for far more costly steps in the decades to come.

Since February 2013, GAO has included *Limiting the Federal Government's Fiscal Exposure by Better Managing Climate Change Risks* on its list of federal program areas at high risk of vulnerabilities to fraud, waste, abuse, and mismanagement or most in need of transformation. GAO updates this list every 2 years. In March 2019, GAO reported that the federal government had not made measurable progress since 2017 to reduce fiscal exposure to climate change.

This testimony—based on reports GAO issued from October 2009 to March 2019—discusses (1) what is known about the potential economic effects of climate change in the United States and the extent to which this information could help federal decision makers manage climate risks across the federal government, (2) the potential impacts of climate change on the federal budget, (3) the extent to which the federal government has invested in resilience, and (4) how the federal government could reduce fiscal exposure to the effects of climate change.

GAO has made 62 recommendations related to the *Limiting the Federal Government's Fiscal Exposure by Better Managing Climate Change Risks* high-risk area. As of December 2018, 25 of those recommendations remained open.

WHAT GAO FOUND

The estimated economic effects of climate change, while imprecise, can convey useful insight about potential damages in the United States. In September 2017, GAO reported that the potential economic effects of climate change could be significant and unevenly distributed across sectors and regions (see figure). This is consistent with the recent findings of the

U.S. Global Change Research Program's Fourth National Climate Assessment, which concluded, among other things, that the continued increase in the frequency and extent of high-tide flooding due to sea level rise threatens America's trillion-dollar coastal infrastructure.

Information about the potential economic effects of climate change could inform decision makers about significant potential damages in different U.S. sectors or regions. According to prior GAO work, this information could help decision makers identify significant climate risks as an initial step toward managing them.

The federal government faces fiscal exposure from climate change risks in several areas, including:

- Disaster aid: due to the rising number of natural disasters and increasing reliance on federal assistance. GAO has previously reported that the federal government does not adequately plan for disaster resilience. GAO has also reported that, due to an artifically low indicator for determining a jursidiction's ability to respond to disasters that was set in 1986, the Federal Emergency Management Agency risks recommending federal assistance for juridisctions that could recover on their own.
- Federal insurance for property and crops: due, in part, to the vulnerability of insured property and crops to climate change impacts. Federal flood and crop insurance programs were not designed to generate sufficient funds to fully cover all losses and expenses. The flood insurance program, for example, was about $21 billion in debt to the Treasury as of April 2019. Further, the Congressional Budget Office estimated in May 2019 that federal crop insurance would cost the federal government an average of about $8 billion annually from 2019 through 2029.
- Operation and management of federal property and lands: due to the hundreds of thousands of federal facilities and millions of acres of land that could be affected by a changing climate and more frequent extreme events. For example, in 2018, Hurricane Michael

devastated Tyndall Air Force Base in Florida, with a preliminary repair estimate of $3 billion.

The federal budget, however, does not generally account for disaster assistance provided by Congress or the long-term impacts of climate change on existing federal infrastructure and programs. GAO has reported that more complete information about fiscal exposure could help policymakers better understand the trade-offs when making spending decisions.

Further, federal investments in resilience to reduce fiscal exposures have been limited. As GAO has reported, enhancing resilience can reduce fiscal exposure by reducing or eliminating long-term risk to people and property from natural hazards. For example, a 2018 interim report by the National Institute of Building Sciences estimated approximate benefits to society in excess of costs for several types of resilience projects. While precise benefits are uncertain, the report estimated that for every dollar invested in designing new buildings to particular design standards, society could accrue benefits amounting to about $11 on average.

The federal government has invested in individual agency efforts that could help build resilience within existing programs or projects. For example, the National Climate Assessment reported that the U.S. military integrates climate risks into its analysis, plans, and programs. In additon, as GAO reported in March 2019, the Disaster Recovery Reform Act of 2018 could improve resilience by allowing the President to set aside a portion of certain grants for pre-disaster mitigation. However, the federal government has not undertaken strategic government-wide planning to manage climate risks.

GAO's March 2019 High-Risk report identified a number of recommendations GAO has made related to fiscal exposure to climate change. The federal government could reduce its fiscal exposure by implementing these recommendations. Among GAO's key government-wide recommendations are:

- Entities within the Executive Office of the President (EOP) should work with partners to establish federal strategic climate change priorities that reflect the full range of climate-related federal activities;
- Entities within EOP should use information on potential economic effects from climate change to help identify significant climate risks and craft appropriate federal responses;
- Entities within EOP should designate a federal entity to develop and update a set of authoritative climate observations and projections for use in federal decision making, and create a national climate information system with defined roles for federal agencies and certain nonfederal entities; and
- The Department of Commerce should convene federal agencies to provide the best-available forward-looking climate information to organizations that develop design standards and building codes to enhance infrastructure resilience.

Chairman Yarmuth, Ranking Member Womack, and Members of the Committee:

Thank you for the opportunity to discuss our work on how to limit the federal government's fiscal exposure by better managing climate change risks, an area that has been on our High-Risk List since February 2013.[1] Addressing climate change risks requires advanced planning and investment to reduce the need for far more costly steps in the decades to come, which, as we have previously reported, the federal government is not well organized to do. The costs associated with recent disasters have illustrated the need for such planning and investment. In 2018 alone, there were 14 separate billion-dollar weather and climate disaster events across the United States, with a total cost of at least $91 billion, according to the

---

[1] Our High-Risk List identifies federal program areas that are at high risk of vulnerabilities to fraud, waste, abuse, and mismanagement or most in need of transformation. See GAO, *High-Risk Series: An Update*, GAO-13-283 (Washington, D.C.: Feb. 14, 2013).

National Oceanic and Atmospheric Administration (NOAA).[2] Further, on June 6, 2019, a supplemental appropriation of approximately $19.1 billion was signed into law for recent disasters.

The U.S. Global Change Research Program (USGCRP), which coordinates and integrates the activities of 13 federal agencies that research changes in the global environment and their implications for society, reported in its November 2018 Fourth National Climate Assessment that climate change is playing a role in the increasing frequency of some types of extreme weather that lead to the billion-dollar disasters.[3] These changes include the rise in vulnerability to drought, lengthening wildfire seasons, and the potential for extremely heavy rainfall becoming more common in some regions. USGCRP reported in the prior assessment that the costs of many of these disasters will likely increase as extreme weather events become more frequent and intense with climate change.[4]

In my testimony today, I will discuss (1) what is known about the potential economic effects of climate change in the United States and the extent to which this information could help federal decision makers manage climate risks across the federal government, (2) the potential impacts of climate change on the federal budget, (3) the extent to which the federal government has invested in resilience to climate change impacts,[5]

---

[2] NOAA National Centers for Environmental Information, U.S. Billion-Dollar Weather and Climate Disasters (2019). See: https://www.ncdc.noaa.gov/billions/time-series, accessed June 3, 2019.

[3] D.R. Reidmiller, C.W. Avery, D. R. Easterling, K. E. Kunkel, K. L. M. Lewis, T. K. Maycock, and B. C. Stewart (eds.), *2018: Impacts, Risks, and Adaptation in the United States: Fourth National Climate Assessment, Volume II* (Washington, DC: U.S. Global Change Research Program, November 2018). Under the Global Change Research Act of 1990 (Pub. L. No. 101-606, § 103 (1990)), USGCRP is to periodically prepare a scientific assessment—known as the National Climate Assessment—which is an important resource for understanding and communicating climate change science and impacts in the United States. The Office of Science and Technology Policy within the Executive Office of the President oversees USGCRP.

[4] Jerry M. Melillo, Terese (T.C.) Richmond, and Gary W. Yohe, eds., *Climate Change Impacts in the United States: The Third National Climate Assessment*, U.S. Global Change Research Program (Washington, D.C.: May 2014).

[5] The National Academies of Sciences, Engineering, and Medicine (National Academies) define resilience as the ability to prepare and plan for, absorb, recover from, and more successfully adapt to adverse events. See the National Academies, Committee on Increasing National Resilience to Hazards and Disasters; Committee on Science, Engineering, and Public Policy; Disaster Resilience: A National Imperative (Washington, D.C.: 2012). We reported

and (4) how the federal government could reduce fiscal exposure to the effects of climate change. My testimony is based on reports we issued from October 2009 to March 2019. More detailed information on our objectives, scope, and methodology can be found in those reports.

The work upon which this statement is based was conducted in accordance with generally accepted government auditing standards. Those standards require that we plan and perform the audit to obtain sufficient, appropriate evidence to provide a reasonable basis for our findings and conclusions based on our audit objectives. We believe that the evidence obtained provides a reasonable basis for our findings and conclusions based on our audit objectives.

Information on the Potential Economic Effects of Climate Change in the United States Could Help Federal Decision Makers Better Manage Climate Risks

We reported in September 2017 that, while estimates of the economic effects of climate change are imprecise due to modeling and information limitations, they can convey useful insight into broad themes about potential damages in the United States.[6] We reported that, according to the two national-scale studies available at the time that examined the economic effects of climate change across U.S. sectors, potential economic effects could be significant and these effects will likely increase over time for

in May 2016 that two related sets of actions can enhance resilience by reducing risk. These include climate change adaptation and pre-disaster hazard mitigation. Adaptation is defined as adjustments to natural or human systems in response to actual or expected climate change. Pre-disaster hazard mitigation refers to actions taken to reduce the loss of life and property by lessening the impacts of adverse events and applies to all hazards, including terrorism and natural hazards, such as health pandemics or weather-related disasters. In this testimony, we use the term "resilience" for consistency and to encompass both of these sets of actions as they relate to addressing climate risks. GAO, *Climate Change: Selected Governments Have Approached Adaptation through Laws and Long-Term Plans,* GAO-16-454 (Washington, D.C.: May 12, 2016).

[6] GAO, *Climate Change: Information on Potential Economic Effects Could Help Guide Federal Efforts to Reduce Fiscal Exposure*, GAO-17-720 (Washington, D.C.: Sept. 28, 2017).

most of the sectors analyzed.[7] For example, for 2020 through 2039, one of the studies estimated from $4 billion to $6 billion in annual coastal property damages from sea level rise and more frequent and intense storms.[8] In addition, the national-scale studies we reviewed and several experts we interviewed for the September 2017 report suggested that potential economic effects could be unevenly distributed across sectors and regions. For example, one of the studies estimated that the Southeast, Midwest, and Great Plains regions will likely experience greater combined economic effects than other regions, largely because of coastal property damage in the Southeast and changes in crop yields in the Midwest and Great Plains (see Figure 1).[9] This is consistent with the findings of the Fourth National Climate Assessment.[10] For example, according to that assessment, the continued increase in the frequency and extent of high-tide flooding due to sea level rise threatens America's trillion-dollar coastal property market and public infrastructure sector.

---

[7] These national-scale studies were the Environmental Protection Agency's *Climate Change Impacts and Risk Analysis*—a summary study of an ongoing EPA project—and the Rhodium Group's *American Climate Prospectus*. See Environmental Protection Agency, Office of Atmospheric Programs, *Climate Change in the United States: Benefits of Global Action*, EPA 430-R-15-001 (Washington, D.C.: 2015). The EPA project on which the summary study was based was coordinated by EPA's Office of Atmospheric Programs—Climate Change Division, with contributions from national laboratories and the academic and private sectors. The detailed methods and results of the project were published in a 2014 special issue of the peer-reviewed journal, *Climatic Change* entitled, "A Multi-Model Framework to Achieve Consistent Evaluation of Climate Change Impacts in the United States." An update to this project was used in the 2018 Fourth National Climate Assessment. Also see Rhodium Group, LLC., *American Climate Prospectus: Economic Risks in the United States* (New York: October 2014). The *American Climate Prospectus* was funded by the Risky Business Project, a project funded by Bloomberg Philanthropies, the Paulsen Institute, and TomKat Charitable Trust; the Skoll Global Threats Fund; and the Rockefeller Family Fund. The Rhodium Group, LLC, a research consultancy and advisory company, coordinated the effort, which involved authors from universities and the private sector. This study was later published by the Columbia University Press in 2015: Trevor Houser et al., *Economic Risks of Climate Change: An American Prospectus* (New York: Columbia University Press, 2015). An update to this analysis was published in *Science* in June 2017: Solomon Hsiang et al. "Estimating Economic Damage from Climate Change in the United States," *Science*, vol. 356 (2017).

[8] Rhodium Group, *American Climate Prospectus*.

[9] Rhodium Group, *American Climate Prospectus*.

[10] D.R. Reidmiller et. al, *Fourth National Climate Assessment, Volume II*.

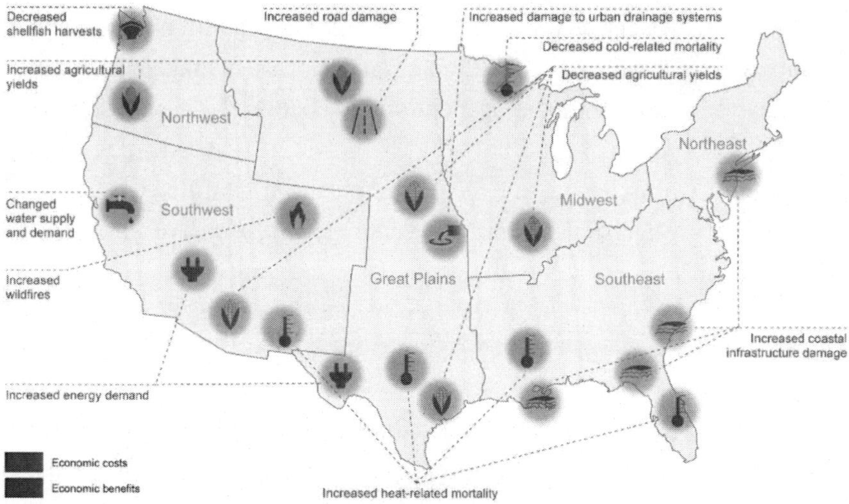

Sources: GAO analysis of Environmental Protection Agency, *Climate Change Impacts in the United States; Benefits of Global Action* (Washington, D.C.: 2015), and Solomon Hsiang et al., "Estimating Economic Damage from Climate Change in the United States," *Science*, vol. 356 (2017); Map Resources (map). | GAO-19-625T.

Figure 1. Examples of Potential Economic Effects from Climate Change by 2100.

As we reported in September 2017, information on the potential economic effects of climate change could help federal decision makers better manage climate risks, according to leading practices for climate risk management, economic analysis we reviewed, and the views of several experts we interviewed.[11] For example, such information could inform decision makers about significant potential damages in different U.S. sectors or regions. According to several experts and our prior work, this information could help federal decision makers identify significant climate priorities as an initial step toward managing climate risks.[12] Such a first step is consistent with leading practices for climate risk management and

---

[11] In that report, we also found that additional economic information could help federal, state, local, and private sector decision makers manage climate risks that drive federal fiscal exposure. GAO-17-720.

[12] GAO-17-720.

federal standards for internal control.[13] For example, leading practices from the National Academies call for climate change risk management efforts that focus on where immediate attention is needed.[14] As noted in our September 2017 report, according to a 2010 National Academies report, other literature we reviewed, and several experts we interviewed, to make informed choices, decision makers need more comprehensive information on economic effects to better understand the potential costs of climate change to society and begin to develop an understanding of the benefits and costs of different options for managing climate risks.[15]

THE FEDERAL GOVERNMENT FACES FISCAL EXPOSURE FROM CLIMATE CHANGE RISKS, BUT DOES NOT HAVE CERTAIN INFORMATION NEEDED TO HELP MAKE BUDGET DECISIONS

The federal government faces fiscal exposure from climate change risks in a number of areas, and this exposure will likely increase over time, as we concluded in September 2017.[16] In the March 2019 update to our High-Risk List, we summarized our previous work that identified several of these areas across the federal government, including programs related to the following:[17]

---

[13] National Research Council of the National Academies, America's Climate Choices: Panel on Adapting to the Impacts of Climate Change, *Adapting to the Impacts of Climate Change* and GAO, *Standards for Internal Control in the Federal Government*, GAO-14-704G (Washington, D.C.: September 2014).

[14] National Research Council of the National Academies, America's Climate Choices: Panel on Adapting to the Impacts of Climate Change, *Adapting to the Impacts of Climate Change* (Washington, D.C.: 2010).

[15] GAO-17-720.

[16] GAO-17-720.

[17] We have identified other areas with potential links to climate and the federal budget in past reports, including global migration, state and local infrastructure, federal supply chains, and public health. See GAO, *Climate Change: Activities of Selected Agencies to Address Potential Impact on Global Migration,* GAO-19-166 (Washington, D.C.: Jan. 17, 2019); *Climate Information: A National System Could Help Federal, State, Local, and Private Sector Decision Makers Use Climate Information,* GAO-16-37 (Washington, D.C.: Nov. 23, 2015); *Federal Supply Chains: Opportunities to Improve the Management of Climate-Related Risks,* GAO-16-32 (Washington, D.C.: Oct. 13, 2015); and *Climate Change: HHS*

- Disaster aid. The rising number of natural disasters and increasing reliance on federal assistance are a key source of federal fiscal exposure, and this exposure will likely continue to rise. Since 2005, federal funding for disaster assistance is at least $450 billion.[18] In September 2018, we reported that four hurricane and wildfire disasters in 2017 created an unprecedented demand for federal disaster resources and that hurricanes Harvey, Irma, and Maria ranked among the top five costliest hurricanes on record.[19] Subsequently, the fall of 2018 brought additional catastrophic disasters such as Hurricanes Florence and Michael and devastating California wildfires, with further needs for federal disaster assistance. Disaster costs are projected to increase as certain extreme weather events become more frequent and intense due to climate change—as observed and projected by USGCRP.[20] In July 2015, we reported that the federal government does not adequately plan for disaster resilience and that most federal funding for hazard mitigation is available after a disaster.[21] In addition, our prior work

---

Could Take Further Steps to Enhance Understanding of Public Health Risks, GAO-16-122 (Washington, D.C.: Oct. 5, 2015). We also have ongoing work in many areas related to federal fiscal exposure to climate change, examining issues such as how to identify and prioritize resilience projects to build resilience to climate change impacts, how to make water infrastructure more resilient to the impacts of climate change, and how to help communities voluntarily relocate to avoid climate change impacts.

[18] This total includes, for fiscal years 2005 through 2014, $278 billion that GAO found that the federal government had obligated for disaster assistance. See GAO, *Federal Disaster Assistance: Federal Departments and Agencies Obligated at Least $277.6 Billion during Fiscal Years 2005 through 2014,* GAO-16-797 (Washington, D.C.: Sept. 22, 2016). It also includes, for fiscal years 2015 through 2018, $124 billion in select supplemental appropriations to federal agencies for disaster assistance, approximately $7 billion in annual appropriations to the Disaster Relief Fund (a total of $28 billion for the 4-year period). For fiscal years 2015 through 2018, it does not include other annual appropriations to federal agencies for disaster assistance. Lastly, on June 6, 2019, the Additional Supplemental Appropriations for Disaster Relief Act of 2019 was signed into law, which provides approximately $19.1 billion for disaster assistance. H.R. 2157, 116th Cong. (2019) (enacted).

[19] GAO, *2017 Hurricanes and Wildfires: Initial Observations on the Federal Response and Key Recovery Challenges*, GAO-18-472 (Washington, D.C.: Sept. 4, 2018).

[20] Jerry M. Melillo, et. al., *Climate Change Impacts in the United States: The Third National Climate Assessment.*

[21] For example, from fiscal years 2011 to 2014, the Federal Emergency Management Agency obligated more than $3.2 billion for the Hazard Mitigation Grant Program for post- disaster hazard mitigation while obligating approximately $222 million for the Pre-Disaster

found that the Federal Emergency Management Agency's (FEMA) indicator for determining whether to recommend that a jurisdiction receive disaster assistance—which was set in 1986—is artificially low because it does not accurately reflect the ability of state and local governments to respond to disasters.[22] Without an accurate assessment of a jurisdiction's capability to respond to a disaster without federal assistance, we found that FEMA runs the risk of recommending that the President award federal assistance to jurisdictions that have the capability to respond and recover on their own.

- Federal insurance for property and crops. The National Flood Insurance Program (NFIP) and the Federal Crop Insurance Corporation are sources of federal fiscal exposure due, in part, to the vulnerability of the insured property and crops to climate change.[23] These programs provide coverage where private markets for insurance do not exist, typically because the risk associated with the property or crops is too great to privately insure at a cost that buyers are willing to accept. From 2013 to 2017, losses paid under NFIP and the federal crop insurance program totaled $51.3 billion.[24] Federal flood and crop insurance programs were not designed to generate sufficient funds to fully cover all losses and expenses, which means the programs need budget authority from Congress to operate. The NFIP, for example, was about $21 billion

Mitigation Grant Program. GAO, *Hurricane Sandy: An Investment Strategy Could Help the Federal Government Enhance National Resilience for Future Disasters*, GAO-15-515 (Washington, D.C.: July 30, 2015).

[22] GAO, *Federal Disaster Assistance: Improved Criteria Needed to Assess a Jurisdiction's Capability to Respond and Recover on Its Own,* GAO-12-838 (Washington, D.C.: Sept. 12, 2012).

[23] The NFIP is administered by FEMA within the U.S. Department of Homeland Security, and the Federal Crop Insurance Corporation is administered by the Risk Management Agency within the U.S. Department of Agriculture.

[24] FEMA and Risk Management Agency published data. This does not include the costs of running these programs or the premiums collected to partially offset the costs. Losses for the crop insurance program are losses associated with crops harvested in that year, also known as crop year.

in debt to the Treasury as of April 2019.[25] Further, the Congressional Budget Office estimated in May 2019 that federal crop insurance would cost the federal government an average of about $8 billion annually from 2019 through 2029.[26]

- Operation and management of federal property and lands. The federal government owns and operates hundreds of thousands of facilities and manages millions of acres of land that could be affected by a changing climate and represent a significant federal fiscal exposure. For example, the Department of Defense (DOD) owns and operates domestic and overseas infrastructure with an estimated replacement value of about $1 trillion. In September 2018, Hurricane Florence damaged Camp Lejeune and other Marine Corps facilities in North Carolina, resulting in a preliminary Marine Corps repair estimate of $3.6 billion. One month later, Hurricane Michael devastated Tyndall Air Force Base in Florida, resulting in a preliminary Air Force repair estimate of $3 billion and upwards of 5 years to complete the work. In addition, we recently reported that the federal government manages about 650 million acres of land in the United States that could be vulnerable to climate change, including the possibility of more frequent and severe droughts and wildfires.[27] Appropriations for federal wildland fire management activities have increased considerably since the 1990s, as we and the Congressional Research Service have reported.[28]

---

[25] U. S. Department of The Treasury, Bureau of the Fiscal Service. *Monthly Treasury Statement. Table 6. Schedule C* (Washington, D.C.: April 2019).

[26] Congressional Budget Office, *CBO's May 2019 Baseline for Farm Programs* (Washington, D.C.: May 2, 2019).

[27] GAO, *Climate Change: Various Adaptation Efforts Are Under Way at Key Natural Resource Management Agencies,* GAO-13-253 (Washington, D.C.: May 31, 2013).

[28] *GAO, Budget Issues: Opportunities to Reduce Federal Fiscal Exposures Through Greater Resilience to Climate Change and Extreme Weather,* GAO-14-504T (Washington, D.C.: July 29, 2014) and Congressional Research Service, *Wildfire Suppression Spending: Background, Issues, and Legislation in the 115th Congress,* R44966 (Washington, D.C.: November 8, 2017).

Although the federal government faces fiscal exposure from climate change across the nation, it does not have certain information needed by policymakers to help understand the budgetary impacts of such exposure.[29] We have previously reported that the federal budget generally does not account for disaster assistance provided by Congress—which can reach tens of billions of dollars for some disasters—or the long-term impacts of climate change on existing federal infrastructure and programs.[30] For Example, as we reported in April 2018, the Office of Management and Budget's (OMB) climate change funding reports we reviewed did not include funding information on federal programs with significant fiscal exposures to climate change identified by OMB and others—such as domestic disaster assistance, flood insurance, and crop insurance.[31] A more complete understanding of climate change fiscal exposures can help policymakers anticipate changes in future spending and enhance control and oversight over federal resources, as we reported in October 2013.[32] For budget decisions for federal programs with fiscal exposure to climate change, we found in the April 2018 report that information that could help provide a more complete understanding would include: (1) costs to repair, replace, and improve the weather- related resilience of federally-funded property and resources; (2) costs for federal flood and crop insurance programs; and (3) costs for disaster assistance programs, among other identified areas of fiscal exposure to climate change.[33] To help policymakers better understand the trade-offs when making spending decisions, we recommended in the April 2018 report that OMB provide

---

[29] In our past work, we identified broad principles for an effective budget process, including that it should (1) provide information about the long-term effects of decisions; (2) provide information necessary to make important trade-offs between spending with long-term benefits and spending with short-term benefits, and (3) provide for accountability and be transparent, among other principles. Further, in October 2013, we reported that incorporating more complete information on fiscal exposures could help meet these principles for an effective budget process. See GAO, *Budget Process: Enforcing Fiscal Choices,* GAO-11-626T (Washington, D.C.: May 4, 2011) and GAO, *Fiscal Exposures: Improving Cost Recognition in the Federal Budget,* GAO-14-28 (Washington, D.C.: Oct. 29, 2013).

[30] GAO-14-505T.

[31] GAO, *Climate Change: Analysis of Reported Federal Funding,* GAO-18-223 (Washington, D.C.: Apr. 30, 2018).

[32] GAO-14-28.

[33] GAO-18-223.

information on fiscal exposures related to climate change in conjunction with future reports on climate change funding.[34]

FEDERAL INVESTMENTS IN RESILIENCE TO CLIMATE CHANGE IMPACTS HAVE BEEN LIMITED

Although the federal government faces fiscal exposure to climate change, its investments in resilience to climate change impacts have been limited. One way to reduce federal fiscal exposure is to enhance resilience by reducing or eliminating long-term risk to people and property from natural hazards. For example, in September 2018 we reported that elevating homes and strengthened building codes in Texas and Florida prevented greater damages during the 2017 hurricane season.[35] In addition, one company participating in a 2014 forum we held on preparing for climate- related risks noted that for every dollar it invested in resilience efforts, the company could prevent $5 in potential losses.[36] Finally, a 2018 interim report by the National Institute of Building Sciences examined a sample of federal grants for hazard mitigation. The report estimated approximate benefits to society (i.e., homeowners, communities, etc.) in excess of costs for several types of resilience projects through the protection of lives and property, and prevention of other losses.[37] For

---

[34] OMB disagreed with this recommendation and has not implemented it, but we continue to believe that the recommendation is valid. GAO-18-223.

[35] Specifically, FEMA officials said Hurricane Harvey demonstrated how prior hazard mitigation projects prevented greater damages (e.g., elevated homes and equipment sustained less damages). FEMA officials said Florida strengthened its building codes for resilience as a result of Hurricanes Andrew in 1992, and Matthew in 2016. GAO-18-472.

[36] GAO, *Highlights of a Forum: Preparing for Climate-Related Risks: Lessons from the Private Sector,* GAO-16-126SP (Washington, D.C.: Nov. 19, 2015).

[37] This chapter examined a narrow sample of hazard mitigation grants awarded by FEMA, the Economic Development Administration, and the Department of Housing and Urban Development from 1993 to 2016 to address various hazards. Extrapolation to a broader set of grants needs to be interpreted in the context of the selected sample. These hazards included fires at the wildland-urban interface (i.e., fires in areas where homes are built near or among lands prone to wildland fire), hurricane- and tornado-force winds, and riverine floods (i.e., floods that occur when river flows exceed the capacity of the river channel). See Multihazard Mitigation Council, a council of the National Institute of Building Sciences,

example, while precise benefits are uncertain, the report estimated that for every grant dollar the federal government spent on resilience projects, over time, society could accrue benefits amounting to the following:

- About $3 on average from projects addressing fire at the wildland urban interface, with most benefits (69 percent) coming from the protection of property (i.e., avoiding property losses).
- About $5 on average from projects to address hurricane and tornado force winds, with most benefits (89 percent) coming from the protection of lives. This includes avoiding deaths, nonfatal injuries, and causes of post-traumatic stress.
- About $7 on average from projects that buy out buildings prone to riverine flooding, with most benefits (65 percent) coming from the protection of property.

The interim report also estimated that society could accrue benefits amounting to about $11 on average for every dollar invested in designing new buildings to meet the 2018 International Building Code and the 2018 International Residential Code—the model building codes developed by the International Code Council—with most benefits (46 percent) coming from the protection of property.[38]

We reported in October 2009 that the federal government's activities to build resilience to climate change were carried out in an ad hoc manner and were not well coordinated across federal agencies.[39]

Natural Hazard Mitigation Saves: 2018 Interim Report (Washington, D.C.: December 2018).

[38] The International Code Council is a member-focused association with over 64,000 members dedicated to developing model codes and standards used in the design, build, and compliance process to construct safe, sustainable, affordable and resilient structures. The report used a baseline of buildings constructed to a prior generation of codes represented by 1990s-era design and National Flood Insurance Program requirements.

[39] GAO, *Climate Change Adaptation: Strategic Federal Planning Could Help Government Officials Make More Informed Decisions,* GAO-10-113 (Washington, D.C.: Oct. 7, 2009).

Federal agencies have included some of these activities within existing programs and operations—a concept known as mainstreaming.

For example, the Fourth National Climate Assessment reported that the U.S. military integrates climate risks into its analysis, plans and programs, with particular attention paid to climate effects on force readiness, military bases, and training ranges.[40] However, according to the Fourth National Climate Assessment, while a significant portion of climate risk can be addressed by mainstreaming, the practice may reduce the visibility of climate resilience relative to dedicated, stand-alone approaches and may prove insufficient to address the full range of climate risks.[41]

In addition, as we reported in March 2019, the Disaster Recovery Reform Act of 2018 (DRRA) was enacted in October 2018, which could improve state and local resilience to disasters. DRRA, among other things, allows the President to set aside, with respect to each major disaster, a percentage of the estimated aggregate amount of certain grants to use for pre-disaster hazard mitigation and makes federal assistance available to state and local governments for building code administration and enforcement.[42] However, it is too early to tell what impact the implementation of the act will have on state and local resilience.

---

[40] Lempert, R., J. Arnold, R. Pulwarty, K. Gordon, K. Greig, C. Hawkins Hoffman, D. Sands, and C. Werrell. 2018. Reducing Risks Through Adaptation Actions. In *Impacts, Risks, and Adaptation in the United States: Fourth National Climate Assessment, Volume II* (Washington, D.C.: U.S. Global Change Research Program, 2018). We also reported in May 2014 that officials from the Office of the Secretary of Defense and the military departments stated that their goal is to address potential climate change impacts and vulnerabilities through existing infrastructure planning processes so that the effects of climate change are considered in the same way other impacts and vulnerabilities—such as force protection—are currently considered. GAO, *Climate Change Adaptation: DOD Can Improve Infrastructure Planning and Processes to Better Account for Potential Impacts,* GAO-14-446 (Washington, D.C.: May 30, 2014).

[41] Lempert, R., J. Arnold, R. Pulwarty, K. Gordon, K. Greig, C. Hawkins Hoffman, D. Sands, and C. Werrell, 2018: Reducing Risks Through Adaptation Actions. *In Impacts, Risks, and Adaptation in the United States: Fourth National Climate Assessment, Volume II* (Washington, D.C.: U.S. Global Change Research Program, 2018).

[42] FAA Reauthorization Act of 2018, Pub. L. No. 115-254, div. D, §§ 1206(a)(3), 1234(a)(2)(C), 1234(a)(5), 132 Stat. 3186, 3440, 3462 (2018). The FAA Reauthorization Act of 2018, which included the DRRA, became law on October 5th, 2018.

The federal government has made some limited investments in resilience and DRRA could enable additional improvements at the state and local level. However, we reported in September 2017 that the federal government had not undertaken strategic government-wide planning to manage significant climate risks before they become fiscal exposures.[43] We also reported in July 2015 that the federal government had no comprehensive strategic approach for identifying, prioritizing, and implementing investments for disaster resilience.[44] As an initial step in managing climate risks, most of the experts we interviewed for the September 2017 report told us that federal decision makers should prioritize risk management efforts on significant climate risks that create the greatest fiscal exposure.[45] However, as we reported in our March 2019 High-Risk List, the federal government had not made measurable progress since 2017 to reduce fiscal exposure in several key areas that we have identified.[46] The High-Risk List identified *Limiting the Federal Government's Fiscal Exposure by Better Managing Climate Change Risks* as an area needing significant attention because the federal government has regressed in progress toward one of our criterion for removal from the list.[47]

---

[43] GAO-17-720.

[44] In our 2015 report, we recommended that the Mitigation Framework Leadership group— an interagency body chaired by FEMA—create a National Mitigation Investment Strategy to help federal, state, and local officials plan for and prioritize disaster resilience. In response, the Mitigation Framework Leadership Group developed a draft, high-level strategy. FEMA officials expect to publish the final version of the strategy by July 2019. However, the draft strategy does not explicitly address future climate change risks. GAO-15-515.

[45] GAO-17-720.

[46] GAO-19-157SP.

[47] We update our High-Risk List every 2 years. To determine which federal government programs and functions should be designated high-risk, we consider qualitative factors such as whether the risk could result in significantly impaired service, or significantly reduced economy, efficiency, or effectiveness; the exposure to loss in monetary or other quantitative terms; and corrective measures planned or under way. We have issued the following five criteria for an area to be removed from the list: leadership commitment, capacity, action plan, monitoring, and demonstrated progress. In the March 2019 report, the federal government regressed in progress toward meeting the monitoring criterion for the *Limiting the Federal Government's Fiscal Exposure by Better Managing Climate Change Risks* high-risk area. Criteria for removing this area from the High-Risk List include demonstrating leadership commitment that is sustained and enhanced to address all aspects of the federal fiscal exposure to climate change cohesively.

THE FEDERAL GOVERNMENT COULD REDUCE ITS FISCAL EXPOSURE BY FOCUSING AND COORDINATING FEDERAL EFFORTS

As we reported in March 2019, the federal government could reduce its fiscal exposure to climate change by focusing and coordinating federal efforts.[48] However, the federal government is currently not well organized to address the fiscal exposure presented by climate change, partly because of the inherently complicated and crosscutting nature of the issue. We have made a total of 62 recommendations related to limiting the federal government's fiscal exposure to climate change over the years, 12 of which have been made since February 2017. As of December 2018, 25 of these recommendations remained open. In describing what needs to be done to reduce federal fiscal exposure to climate change, our March 2019 High-Risk report discusses many of the open recommendations.[49] Implementing these recommendations could help reduce federal fiscal exposure. Several of them, including those highlighted below, identify key government-wide efforts needed to help plan for and manage climate risks and direct federal efforts toward common goals, such as improving resilience:

- Develop a national strategic plan: In May 2011, we recommended that appropriate entities within the Executive Office of the President (EOP), including OMB, work with agencies and interagency coordinating bodies to establish federal strategic climate change priorities that reflect the full range of climate-related federal activities, including roles and responsibilities of key federal entities.[50]
- Use economic information to identify and respond to significant climate risks: In September 2017, we recommended that the

[48] GAO-19-157SP.

[49] GAO-19-157SP.

[50] EOP neither agreed nor disagreed with our recommendation and as of March 2019, had not implemented it. GAO, *Climate Change: Improvements Needed to Clarify National Priorities and Better Align Them with Federal Funding Decisions,* GAO-11-317 (Washington, D.C.: May 20, 2011).

appropriate entities within EOP use information on the potential economic effects of climate change to help identify significant climate risks facing the federal government and craft appropriate federal responses.[51] Such federal responses could include establishing a strategy to identify, prioritize, and guide federal investments to enhance resilience against future disasters.

- Provide decision makers with the best available climate information: In November 2015, we reported that federal efforts to provide information about climate change impacts did not fully meet the climate information needs of federal, state, local, and private sector decision makers, which hindered their efforts to plan for climate change risks.[52] We reported that these decision makers would benefit from a national climate information system that would develop and update authoritative climate observations and projections specifically for use in decision-making. As a result, we recommended that EOP (1) designate a federal entity to develop and periodically update a set of authoritative climate observations and projections for use in federal decision-making, which other decision makers could also access; and (2) designate a federal entity to create a national climate information system with defined roles for federal agencies and nonfederal entities with existing statutory authority.[53]
- Consider climate information in design standards: In November 2016, we reported that design standards, building codes, and voluntary certifications established by standards-developing organizations play a role in ensuring the resilience of infrastructure to the effects of natural disasters. However, we reported that these organizations faced challenges to using forward-looking climate information that could help enhance the resilience of infrastructure. As a result, we recommended in the November 2016 report that the

[51] EOP neither agreed nor disagreed with this recommendation and as of March 2019, had not implemented it. GAO-17-720.

[52] GAO-16-37.

[53] EOP neither agreed nor disagreed with these recommendations and as of March 2019, had not implemented them.

Department of Commerce, acting through the National Institute of Standards and Technology—which is responsible for coordinating federal participation in standards organizations—convene federal agencies for an ongoing government-wide effort to provide the best available forward-looking climate information to standards-developing organizations for their consideration in the development of design standards, building codes, and voluntary certifications.[54]

In conclusion, the effects of climate change have already and will continue to pose risks that can create fiscal exposure across the federal government and this exposure will continue to increase. The federal government does not generally account for such fiscal exposure to programs in the budget process nor has it undertaken strategic efforts to manage significant climate risks that could reduce the need for far more costly steps in the decades to come. To reduce its fiscal exposure, the federal government needs a cohesive strategic approach with strong leadership and the authority to manage risks across the entire range of related federal activities. The federal government could make further progress toward reducing fiscal exposure by implementing the recommendations we have made.

Chairman Yarmuth, Ranking Member Womack, and Members of the Committee, this completes my prepared statement. I would be pleased to respond to any questions that you may have at this time.

---

[54] Commerce neither agreed nor disagreed with this recommendation and as of May 2018, had not implemented it. GAO, *Climate Change: Improved Federal Coordination Could Facilitate Use of Forward-Looking Climate Information in Design Standards, Building Codes, and Certifications,* GAO-17-3 (Washington, D.C.: Nov. 30, 2016).

In: Climate Change
Editor: Jeremy Nesbitt
ISBN: 978-1-53617-216-4

Chapter 5

THE ENDANGERED SPECIES ACT AND CLIMATE CHANGE: SELECTED LEGAL ISSUES*

Linda Tsang

SUMMARY

For more than a decade, federal agencies have grappled with how to address climate change effects when implementing the Endangered Species Act of 1973 (ESA). The ESA aims to protect threatened and endangered fish, wildlife, and plants from extinction. As set forth by Congress, one of the main purposes of the ESA is to "provide a means whereby the ecosystems upon which endangered species and threatened species depend may be conserved."

The U.S. Fish and Wildlife Service (FWS) and the National Marine Fisheries Service (NMFS) (collectively, the Services) have acknowledged that the changing climate may threaten the survival of and habitat for some species. As noted by courts and legal scholars, the ESA does not expressly require the Services to consider the effect of climate change in their ESA decisions. However, the ESA and its implementing regulations (1) direct the Services to consider "natural or manmade factors affecting

* This is an edited, reformatted and augmented version of Congressional Research Service, Publication No. R45926, dated September 20, 2019.

[a species'] continued existence" when determining whether a species should be protected under the ESA; and (2) require the Services to analyze cumulative effects on a species' survival when analyzing whether federal actions jeopardize a species protected under the Act. The courts and the Services have interpreted these provisions as requiring the Services to consider climate change effects in the ESA decisionmaking process. Various lawsuits have challenged the Services' interpretation of complex scientific data or models that predict short- and long-term effects from a changing global climate on specific species and their habitats.

Legal challenges have influenced how the Services implement the ESA when climate change affects species and their habitats. Lawsuits typically focus on two main issues: (1) when the Services should list, delist, or reclassify a species as threatened or endangered because of climate change effects; and (2) whether the Services can or should regulate activities that affect the climate to protect species and their habitat. Judicial review has helped to ensure that the Services consider projected climate change effects on species in their ESA decisions. However, the courts have not required the Services to curb activities that may contribute to climate change to protect threatened or endangered species.

Stakeholders disagree on whether the ESA should play a role in addressing climate change, with some arguing that the ESA is not equipped to mitigate climate change effects. Other stakeholders believe that the Services can and should wield the ESA to protect further species threatened by climate change by curbing activities contributing to climate change. From the Services' viewpoint, the best available scientific and commercial data have been insufficient to determine whether greenhouse gas emissions from a proposed activity cause detrimental effects on a species or its habitat. In light of the judicial deference afforded to the Services, the courts have not expanded the ESA as a tool to protect listed species by regulating activities that contribute to climate change.

This chapter analyzes the courts' role in shaping how the Services have factored climate change effects into ESA decisions and recent 2019 regulatory developments that aim to clarify how the Services consider and address climate change in their ESA decisions. In August 2019, the Services finalized revisions to the ESA implementing regulations, aiming to increase transparency and effectiveness of the ESA while easing regulatory burdens. Among those changes, the Services clarified their existing policies and practices for factoring climate change effects into their ESA decisions. As legislative proposals to revise the ESA continue to develop, legal battles over the how the Services interpret climate change effects in their ESA decisions will likely continue.

For more than a decade, federal agencies have grappled with how to address climate change effects when implementing the Endangered Species Act of 1973 (ESA, or the Act). The ESA aims to protect threatened and endangered fish, wildlife, and plants from extinction.[1]

As set forth by Congress, one of the main purposes of the ESA is to "provide a means whereby the ecosystems upon which endangered species and threatened species depend may be conserved."[2] To conserve threatened and endangered species, the Act seeks to identify threatened or endangered species, facilitate recovery and conservation of these species, and minimize the effect of federal and private actions on these species and their habitats.[3] The Supreme Court has stated that "[t]he plain intent of Congress in enacting this statute was to reverse the trend toward species extinction, whatever the cost."[4]

To achieve that purpose, Congress declared that "all Federal departments and agencies shall seek to conserve endangered species and threatened species and shall utilize their authorities" to further the ESA purposes.[5] Under the ESA, two federal agencies—the U.S. Fish and Wildlife Service (FWS) within the Department of the Interior and the National Marine Fisheries Service (NMFS) within the Department of Commerce (collectively, the Services)—are primarily responsible for implementing the ESA.[6] According to the Services, over 1,500 species of plants and animals receive some type of protection under the ESA.[7] Since

---

[1] P.L. 93-205, 87 Stat. 884. 16 U.S.C. §§1531-1544. For more information on the ESA, see CRS In Focus IF11241, *The Legal Framework of the Endangered Species Act (ESA)*, by Erin H. Ward; and CRS Report RL31654, *The Endangered Species Act: A Primer*, by Pervaze A. Sheikh.

[2] 16 U.S.C. § 1531(b).

[3] For a description of the legal framework of the ESA, see CRS In Focus IF11241, *The Legal Framework of the Endangered Species Act (ESA)*, by Erin H. Ward.

[4] Tennessee Valley Authority v. Hill, 437 U.S. 153, 184 (1978).

[5] 16 U.S.C. § 1531(c)(1).

[6] In general, FWS manages terrestrial, freshwater, and catadromous species, and NMFS manages marine species and anadromous fish. For detailed information on FWS's ESA program, see FWS website at http://www.fws.gov/ endangered/. For detailed information on NMFS's ESA program, see NMFS website at http://www.nmfs.noaa.gov/pr/ laws/esa/. NMFS, a part of the National Oceanic and Atmospheric Administration (NOAA), is also referred to as *NOAA Fisheries*.

[7] *Environmental Conservation Online System, Listed Animals*, FISH & WILDLIFE SERV., https://go.usa.gov/xVXNs (last visited Sept. 9, 2019); *Species Directory*, NOAA

the early 21st century, some Members of Congress have urged the Services to factor in climate change effects when implementing the ESA.[8]

The Services, along with scholars and scientists, have acknowledged that the changing climate may threaten the survival of and habitat for some species.[9] As noted by courts and legal scholars, the ESA does not expressly require the Services to consider the effect of climate change in their ESA decisions.[10] However, the ESA and its implementing regulations (1) direct the Services to consider "natural or manmade factors affecting [a species'] continued existence" when determining whether a species should be protected under the ESA; and (2) require the Services to analyze cumulative effects on a species' survival when analyzing whether federal actions jeopardize a species protected under the Act.[11] The courts and the

FISHERIES, https://www.fisheries.noaa.gov/species- directory/threatened-endangered (last visited Sept. 19, 2019).

[8] *See, e.g.*, H.R. Rep. No. 110-296, Pt. 1, at 35 (2007) (allocating funds to FWS endangered species programs to assist species adaptation to climate change). *See also* Dean Scott, *Appropriators Urge Interior to Deepen Review of How Global Warming is Affecting Species*, DAILY ENV'T REP. (BNA) (Apr. 27, 2007) (discussing congressional hearing urging climate change research to determine whether and how species are being affected).

[9] *See* Colo. River Cutthroat Trout v. Salazar, 898 F. Supp. 2d 191, 207 (D.D.C. 2012) ("As scientific assessments increasingly incorporate in-depth analyses of climate change effects, explicit consideration of climate change-related threats may become a necessary component of the status review."); NMFS, REVISED GUIDANCE FOR TREATMENT OF CLIMATE CHANGE IN NMFS ENDANGERED SPECIES ACT DECISIONS 1 (2016) ("[C]limate change has become a key lens through which resource management decisions must be evaluated and addressed."). *See also* David Owen, *Endangered Species Act, in* GLOBAL CLIMATE CHANGE AND U.S. LAW 188 (Michael B. Gerrard & Jody Freeman eds., 2d ed. 2014); James Ming Chen, *The Fragile Menagerie: Biodiversity Loss, Climate Change, and the Law*, 93 IND. L.J. 303, 304-310 (2018); Barry Kellman, *Climate Change in the Endangered Species Act: A Jurisprudential Enigma*, 46 ENVTL. L. REP. 10845, 10845 (2016).

[10] *See* Colo. River Cutthroat Trout v. Salazar, 898 F. Supp. 2d 191, 206-07 (D.D.C. 2012) (noting "no statutory requirement" requiring the Services "to discuss climate change in [their] listing decisions"); All. for the Wild Rockies v. Lyder, 728 F. Supp. 2d 1126, 1140 (D. Mont. 2010) (upholding FWS's decision to not designate critical habitat for the Canada link based only on regional climate data). *See also* James Ming Chen, *The Fragile Menagerie: Biodiversity Loss, Climate Change, and the Law*, 93 IND. L.J. 303, 334 (2018).

[11] 16 U.S.C. § 1533(a)(1); 50 C.F.R. § 402.14(g); Pac. Coast Fed'n of Fishermen's Ass'ns v. Gutierrez, 606 F. Supp. 2d 1122, 1184 (E.D. Cal. 2008); Nat. Res. Def. Council v. Kempthorne, 506 F. Supp. 2 322, 374-76 (E.D. Cal. 2007).

Courts had generally found that FWS met the requirement to consider cumulative threats from climate change when it provided "even a brief discussion" of such threats. Desert Survivors v. Dept. of the Interior, 321 F. Supp. 3d 1011, 1049 (N.D. Cal. 2018) (concluding that FWS "offered sufficient explanation of its consideration of cumulative threats" to the bi-state

Services have interpreted these provisions as requiring the Services to consider climate change effects into the ESA decisionmaking process.[12]

Various lawsuits have challenged the Services' interpretation of complex scientific data or models that predict short- and long-term effects from a changing global climate on specific species and their habitats.[13] These lawsuits typically focus on two main issues: (1) when the Services should list, delist, or reclassify a species as threatened or endangered because of climate change effects; and (2) whether the Services can or should regulate activities that affect the climate to protect the species. Judicial review has helped to ensure that the Services consider projected climate change effects on species in their ESA decisions, but the courts have not required the Services to curb activities that may contribute to climate change to protect threatened or endangered species.

This chapter analyzes the courts' role in shaping how the Services have factored climate change effects into ESA decisions and recent regulatory developments that seek to clarify how the Services consider and address climate change in their ESA decisions.

JUDICIAL REVIEW UNDER THE ESA

In general, stakeholders challenge the Services' ESA actions or inactions under the Administrative Procedure Act (APA). The APA authorizes reviewing courts to "hold unlawful and set aside agency actions, findings, and conclusions found to be arbitrary, capricious, [or] an abuse of discretion."[14] Under the arbitrary and capricious standard, courts must determine whether the agency "examine[d] the relevant data and articulate[d] a satisfactory explanation for its action, including a 'rational

---

grouse by "identif[ying] the threats that may interact and provid[ing] some explanation of the implications of the interactions.").

[12] *See infra* "Listing Decisions Under the ESA."

[13] LAWRENCE R. LIEBESMAN & RAFE PETERSEN, ENDANGERED SPECIES DESKBOOK 104 (2010); John Kostyack & Dan Rohlf, *Conserving Endangered Species in an Era of Global Warming, in* ENDANGERED SPECIES: LAW POLICY AND PERSPECTIVE 375, 379 (Donald C. Baur et al. eds. 2010).

[14] 5 U.S.C. § 706(2)(A).

connection between the facts found and the choice made,'" but the standard prohibits courts from "substitut[ing] its judgment for that of the agency."[15] Under this deferential standard, courts have generally deferred to the Services' decisions related to climate change.

However, courts have not deferred to the Services when the court concludes that the record does not support the Services' decision or the Services failed to consider climate change adequately.[16]

The sections below offer selected examples, drawn from various court decisions, legal documents, and regulatory developments, to illustrate the range of issues that the Services and the courts have addressed related to the ESA and climate change. Each section of the report reviews the applicable legal framework and discusses the relevant regulatory revisions finalized by the Trump Administration in August 2019. This chapter does not aim to provide a comprehensive or representative preview of all the judicial decisions that have addressed this area.

LISTING DECISIONS UNDER THE ESA

Many legal challenges involving the ESA and climate change have centered on whether to list a species as endangered or threatened under the ESA. To trigger protections and prohibitions under the ESA, the Services must first list a species as threatened or endangered. Under ESA Section 4, the Services list a species as endangered or threatened based on assessments of the risk of their extinction.[17] The Act defines an "endangered species" as a species "in danger of extinction throughout all or a significant portion of its range."[18] A "threatened species" is a species "likely to become endangered within the foreseeable future in all or a

---

[15] Motor Vehicle Mfrs. Ass'n of U.S. v. State Farm Mut. Auto Ins. Co., 463 U.S. 29, 43 (1983) (quoting Burlington Truck Lines, Inc. v. United States 371 U.S. 156, 168 (1962)).

[16] Barry Kellman, *Climate Change in the Endangered Species Act: A Jurisprudential Enigma*, 46 ENTVL. L. REP. 10845, 10846 (2016).

[17] 16 U.S.C. §1533(a).

[18] *Id.* § 1532(6).

significant portion of its range."[19] For listing decisions, the ESA requires the Services to determine whether the species "is a threatened or endangered species because of any of the following factors:

(A) the present or threatened destruction, modification, or curtailment of its habitat or range;
(B) overutilization for commercial, recreational, scientific, or educational purposes;
(C) disease or predation;
(D) the inadequacy of existing regulatory mechanisms; or
(E) other natural or manmade factors affecting its continued existence."[20]

When listing a species, the Services must make their decision "solely on the basis of the best scientific and commercial data available . . . after conducting a review of the status of the species," taking into account any state's or foreign nation's actions to protect such species.[21]

Courts have consistently held that the Services must consider climate change as a factor in their listing decisions if it may affect the survival of the species.[22] However, stakeholders have disputed the extent to which climate change affects species and the science underpinning listing decisions. Some stakeholders have sought through petitions[23] and legal challenges to compel the Services to list species whose survival has been or may be threatened by climate change effects.[24] Other stakeholders have

---

[19] *Id.* § 1532(20).
[20] *Id.* § 1533(a)(1).
[21] *Id.* § 1533(b)(1)(A).
[22] *See, e.g.*, Greater Yellowstone Coalition v. Servheen, 665 F.3d 1015 (9th Cir. 2011).
[23] Interested parties may petition the Services to list a species as endangered or threatened under the ESA. 16 U.S.C. § 1533(b)(2).
[24] *See, e.g.*, Animal Welfare Inst. & Def. of Wildlife, *Petition to List the Northwest Atlantic Distinct Population Segment (DPS) of the Thorny Skate (*Amblyraja radiata*) as Endangered or Threatened or, Alternatively, to List the United States DPS of the Thorny Skate as Endangered or Threatened under the U.S. Endangered Species Act* (May 28, 2015), https://awionline.org/sites/default/files/uploads/documents/AWI-ML-ThornySkate-ESA-Petition-submitted.pdf; Ctr. For Biological Diversity, *Petition to List the Pacific Walrus (*Odobenus rosmarus divergens*) as Threatened or Endangered under the Endangered Species Act* (Feb. 7, 2008),

challenged the listing of species or petitioned the Service to delist a species, questioning whether model-based climate predictions constitute the "best scientific and commercial data available" on which to base ESA listing decisions.[25]

Scientific uncertainty and undefined terms in the ESA have opened the door to litigation challenging the Services' interpretation of ambiguous terms and their assessment of the climate science that supports their listing decisions. Courts often uphold the Services' interpretation of ambiguous terms because judicial review of agency decisions is narrow and highly deferential; the court will not set aside an ESA listing decision so long as it is rational and reasonably based on supporting evidence.[26] However, courts have faulted the Services for inadequately considering climate change effects or relying on the scientific uncertainty of climate modeling to deny a petition to list a species.[27] The two sections below discuss various court decisions that have reviewed how the Services (1) interpret the undefined "foreseeable future" in their listing decisions, and (2) address the scientific uncertainty of climate change effects.

Foreseeability of Climate Change Effects in Listing Decisions

Legal challenges to Services' decisions to list or not to list a species as threatened highlight the difficulty in predicting whether a species is likely to be endangered "within the foreseeable future" because of climate change effects.[28] Neither the ESA nor the implementing regulations define the term *foreseeable future*. Under their interpretation of the term, the Services determine foreseeability on a case-by-case basis for listing decisions, and the foreseeable future time frame can vary considerably based on the

https://www.biologicaldiversity.org/species/mammals/Pacific_walrus/pdfs/CBD-Pacific-walrus-petition.pdf.

[25] *See infra* "Foreseeability of Climate Change Effects in Listing Decisions."

[26] *See* Alaska Oil & Gas Ass'n v. Jewell, 815 F.3d 544, 554 (9th Cir. 2016).

[27] *See infra* "Scientific Uncertainty in Listing Decisions."

[28] 16 U.S.C. § 1532(20).

species and its habitat.[29] For species affected by climate change, the Services' decisions on foreseeability of a species' survival often depend on their assessment of predictive modeling of climate threats to a species and its habitat. How a Service defines a species' foreseeable future could affect its ESA listing decision. For example, a species is less likely to be listed for protection under the ESA if the Services adopt a shorter time frame for the foreseeable future, thereby limiting their consideration of longer-term projections of climate change effects on a species and its habitat.[30]

The legal challenges to FWS's listing of the polar bear (*Ursus maritimus*) illustrate how courts have applied this narrow and deferential standard of review and interpreted the ESA standards for the best available data in the climate change context. In 2013, in *Safari Club International v.*

Salazar (In re Polar Bear Endangered Species Act Listing and Section 4(d) Rule Litigation) (hereinafter *In re Polar Bear*), the U.S. Court of Appeals for the District of Columbia Circuit (D.C. Circuit) upheld FWS's listing of polar bears as a threatened species under the ESA based in part on projected climate change effects to the species and its habitat.[31] FWS based its decision on three main conclusions: (1) that the polar bear is dependent on sea ice for its survival; (2) that sea ice is declining; and (3)

---

[29] Ctr. for Biological Diversity v. Lubchenco, 758 F. Supp. 2d 945, 967 (N.D. Cal. 2010) ("Plaintiffs concede that the length of time that constitutes the 'foreseeable future' for listing purposes may vary depending on the species and the threats it faces."). *See also* Memorandum (M-37021) from the David L. Bernhardt, Solicitor, Dep't of the Interior, to Acting Director, FWS (Jan. 16, 2009) (providing guidance to FWS on addressing the concept of the foreseeable future in determining whether a species is threatened under the ESA), https://www.fws.gov/endangered/esa-library/pdf/M-37021%20Foreseeable%20 future.pdf.

[30] *See* Jake Li & Angus McLean., *Why for the Forseeable Future Matters*, ENVTL. POLICY & INNOVATION CENTER, http://policyinnovation.org/foreseeable-future (last visited September 19, 2019) (analyzing how agencies have interpreted "foreseeable future" in ESA listing decisions).

[31] 709 F.3d 1 (D.C. Cir. 2013), *cert. denied*, 571 U.S. 887 (2013). Note that the lower court had upheld FWS's listing against an array of consolidated legal challenges that had argued that the polar bear should have been listed as "endangered" rather than "threatened," as well as from industry groups and other plaintiffs that had argued that the polar bear listing was unwarranted. *See In re Polar Bear Endangered Species Act Listing and §4(d) Rule Litigation)*, 838 F. Supp. 2d 214, 218 (D.D.C. 2011). The lower court had ordered FWS to conduct additional environmental review of the ESA Section 4(d) rule for polar bears, *Id.* at 238, which FWS did in 2013. *See* 78 Fed. Reg. 11,766 (Feb. 20, 2013).

that climate change will likely continue to reduce the extent and quality of arctic sea ice gravely enough to endanger the polar bear population.[32]

The D.C. Circuit held that the challenges to FWS's scientific assessment and conclusions "'amount to nothing more than competing views about policy and science,' on which we defer to the agency."[33] The court also rejected arguments that climate science was too uncertain to support listing the polar bear as a species that is likely to become endangered in the "foreseeable future," defined by FWS in this case as 45 years in the future.[34] The court concluded that FWS's reliance on climate projections was "justifiable[,] clearly articulated[,] . . . sufficient to support their definition of foreseeability."[35] The Supreme Court declined to review the case.[36]

In 2016, the U.S. Court of Appeals for the Ninth Circuit (Ninth Circuit) similarly deferred to the NMFS's foreseeable future analysis in upholding the listing of two populations of Arctic bearded seals (*Erignathus barbatus nauticus*) in *Alaska Oil & Gas Association v. Pritzker*.[37] NMFS listed the seals as threatened in 2012 based on climate change models that predicted that sea ice the seals depend on for birthing and mating would mostly disappear by 2095.[38] In rejecting the plaintiffs' claim that the models used in the listing decision could not reliably predict climate change effects on the seals beyond 2050,[39] the Ninth Circuit concluded that NMFS may base its listing decision on such models and long-term projections because the record included a reasonable explanation for its decision.[40] The court explained that the

---

[32] *Id.* at 8.

[33] *Id.* at 9 (quoting *In re Polar Bear Endangered Species Act Listing and §4(d) Rule Litigation*, 794 F. Supp. 2d 65, 69 (D.D.C. 2011)).

[34] *Id.* at 15, quoting 16 U.S.C. §1532(20).

[35] *Id.* at 16.

[36] *Safari Club Int'l v. Jewell*, 571 U.S. 887 (2013).

[37] Alaska Oil & Gas Ass'n v. Pritzker, 840 F.3d 671 (9th Cir. 2016), *cert. denied*, 138 S. Ct. 924 (2018).

[38] *Id.* at 681-82.

[39] *Id.* at 681.

[40] *Id.*

> ESA does not require NMFS to base its decision on ironclad evidence when it determines that a species is likely to become endangered in the foreseeable future; it simply requires the agency to consider the best and most reliable scientific and commercial data and to identify the limits of that data when making a listing determination.[41]

Soon after the bearded seal decision, the Ninth Circuit reversed an Alaska federal district court's decision to vacate NMFS's decision to list the Arctic subspecies of the ringed seal (*Phoca hispida hispida*) as threatened under the ESA.[42] Bound by the precedent set in *Pritzker*, the Ninth Circuit concluded that the district court erred when it required more "definitive quantitative data about the Arctic ringed seal population and an extinction threshold" to list the species as threatened under the ESA.[43] The court determined that NMFS's reliance on climate change models that project until 2100 was not arbitrary or capricious because NMFS "provided a reasonable and scientifically supported methodology for addressing volatility in its long-term climate projections, and it represented fairly the shortcomings of those projections—that is all the ESArequires."[44]

Courts have also deferred to NMFS's decisions not to list species when it reasonably demonstrated that long-term predictive climate models were unreliable to support a listing decision. For example, a federal district court in California upheld NMFS's decision not to list the ribbon seal (*Histriophoca fasciata*) as threatened or endangered despite a "likely" population decline related to sea ice loss and ocean acidification.[45] The court held that NMFS reasonably relied on a 40-year time horizon, from 2010 to 2050, to project negative effects from climate change on the sea ice habitat because it determined the models beyond 2050 were unreliable.[46] The court deferred to NMFS's expertise in upholding NMFS's determination that, based on this time frame, the ribbon seal was not likely to become endangered or in danger of extinction in the foreseeable future because the seal is resilient and adaptable to climate change effects on its

---

[41] *Id.*

[42] Alaska Oil & Gas Ass'n v. Ross, 722 F. App'x 666, 668 (9th Cir. 2018).

[43] *Id.* at 668.

[44] *Id.* at 669, quoting *Pritzker*, 840 F.3d at 680 (internal quotations omitted).

[45] Ctr. for Biological Diversity v. Lubchenco, 758 F. Supp. 2d 945, 967 (N.D. Cal. 2010).

[46] *Id.* at 963-67.

habitat.[47] The court concluded that NMFS did not err when it determined that climate models after 2050 were "unreliable" and "too divergent" to use in assessing future threats to the ribbon seal.[48] NMFS determined that the climate models were "too heavily dependent" on estimated greenhouse gas (GHG) emissions from different types of future regulatory controls.[49]

These foreseeability cases highlight the courts' willingness to defer to the Services' interpretation of climate modeling data and the foreseeability of climate change effects if the record for the listing decision includes a reasonable explanation for their decision that acknowledges limits or uncertainty in the data. As such, the Services continue to evaluate the foreseeability on a case-by- case basis.

Defining "Foreseeable Future" in ESA Regulations

In August 2019, the Services finalized revisions to the ESA regulations to define the "foreseeable future" as extending "only so far into the future as the Services can reasonably determine that both the future threats and the species' responses to those threats are likely."[50] Prior to this final rule, neither the ESA nor the implementing regulations defined the term *foreseeable future*. In the final rule, the Services emphasized that it would continue to evaluate the range of uncertainty and probabilities associated with the best available science and projected data on climate change effects to individual species and their habitat.[51]

It is unclear whether these changes will (1) affect how the Services evaluate long-term projections of climate change effects on species, or (2) promote greater uniformity and consistency within and between FWS and

---

[47] *Id.* at 952-53.
[48] *Id.* at 963-65.
[49] *Id.* at 965.
[50] Endangered and Threatened Wildlife and Plants; Regulations for Listing Species and Designating Critical Habitat, Final Rule, 84 Fed. Reg. 45,020, 45,052 (Aug. 27, 2019).
[51] Endangered and Threatened Wildlife and Plants; Regulations for Listing Species and Designating Critical Habitat, Final Rule, 84 Fed. Reg. 45,020, 45,031-32 (Aug. 27, 2019). The final rule reiterated that the Services would not define foreseeable future in terms of a specific time period and will continue to identify the foreseeable future on a "case-by- case basis, using the best available data and taking into account considerations such as the species' life-history characteristics, threat-projection timeframes, and environmental variability." *Id.* at 45,020.

NMFS in their listing evaluations. Some stakeholders noted that the final rule merely codified the Services' existing practice in determining the foreseeable future for species.[52] Other stakeholders expressed concerns that this definition of *foreseeable future* would limit consideration of long-term projected threats from climate change.[53] In their lawsuit challenging the final rule, plaintiffs claim that demanding that both threats and responses to threats be "likely" in the foreseeable future imposes an "increased certainty requirement" that will deny protection under the ESA for species from the future effects of climate change.[54]

Scientific Uncertainty in Listing Decisions

The legal challenges to the Services' foreseeable future determinations highlight how scientific uncertainty plays a large role in evaluating climate change effects. Similar to the foreseeability cases, courts have faulted the Services for claiming scientific uncertainty without adequate explanation when declining to list a species. This section discusses some examples where stakeholders have challenged FWS's approach to scientific uncertainty in its decisions to not list a species or delist a species under the ESA.

To delist a species under the ESA, the Services must determine that none of the five factors considered in listing the species (i.e., destruction or modification of its habitat or range; overutilization for commercial,

---

[52] Rebecca Barho & Brooke Wahlberg, *Expert Analysis ESA Rule Changes Less Drastic Than Critics Claim,* LAW360 (Aug. 29, 2019), https://www.law360.com/transportation/articles/1192657/esa-rule-changes-less-drastic-than-critics- claim; Jake Li, *Last Week's Endangered Species Regulations: What Really Happened?* ENVTL. POLICY & INNOVATION CENTER (Aug. 20, 2019), http://policyinnovation.org/the-real-story-on-the-new-endangered-species-act-rules/.

[53] *See, e.g.*, Rebecca Beitsch, *Democrats, Environmentalists Blast Trump Rollback of Endangered Species Protections*, THE HILL (Aug. 12, 2019), https://thehill.com/policy/energy-environment/457156-dems-environmentalists-blast-trump-rollback-of-endangered-species; Adam Aton, *Trump Admin Rolls Out Rule Changes to Limit Law's Reach*, GREENWIRE (Aug. 12, 2019), https://www.eenews.net/stories/1060931003

[54] Complaint for Declaratory and Injunctive Relief, Center for Biological Diversity v. Bernhardt, No. 3:19-cv-05206 (N.D. Cal. Aug. 21, 2019).

recreational, scientific, or educational purposes; disease or predation; inadequate existing regulatory protections; and other factors affecting its continued existence) threatens or endangers the species.[55] Delisting determinations must be made "solely on the basis of the best available scientific and commercial information regarding a species' status, without reference to possible economic or other impacts of such determination."[56]

Similar to judicial review of listing decisions discussed above, courts have generally deferred to FWS's decisions regarding scientific uncertainty of climate data unless FWS fails to justify why such uncertainty supports its listing decision. For example, in 2011, the Ninth Circuit in *Greater Yellowstone Coalition Inc. v. Servheen* vacated and remanded FWS's delisting of the Yellowstone grizzly bear (*Ursus arctos horribilis*) as a threatened species, partly because FWS failed to justify why declines in whitebark pine—a primary source of food for grizzlies—due to climate change were not likely to threaten the Yellowstone grizzly bear population.[57] While acknowledging that courts generally defer to the Services' expertise, the Ninth Circuit refused to defer to FWS's "arbitrary" and unsupported claims of scientific uncertainty regarding the effect that declining food supplies resulting from climate change may have on grizzly bears.[58] Relying on evidence that climate change reduced available whitebark pine seeds, increased grizzly bear mortality, and decreased grizzly bear reproduction, the court concluded that overall declines in the grizzly bear's food source from climate change effects in the Yellowstone

---

[55] 50 C.F.R. § 424.11(d).

[56] *Id.* § 424.11(b).

[57] Greater Yellowstone Coal., Inc. v. Servheen, 665 F.3d 1015, 1026 (9th Cir. 2011). In 1975, FWS listed the Yellowstone grizzly bear as threatened under the ESA. Amendment Listing the Grizzly Bear of the 48 Coterminous States as a Threatened Species, 40 Fed. Reg. 31,734 (July 28, 1975). After implementing a multi-agency recovery plan and researching the recovery of grizzly bears, FWS delisted the Yellowstone distinct population segment of grizzly bears from the ESA threatened species list. Endangered and Threatened Wildlife and Plants; Final Rule Designating the Greater Yellowstone Area Population of Grizzly Bears as a Distinct Population Segment; Removing the Yellowstone Distinct Population Segment of Grizzly Bears From the Federal List of Endangered and Threatened Wildlife; 90-Day Finding on a Petition To List as Endangered the Yellowstone Distinct Population Segment of Grizzly Bears, 72 Fed.
Reg. 14,866 (Mar. 29, 2007).

[58] *Greater Yellowstone Coal., Inc.*, 665 F.3d at 1028.

region would logically have a "negative effect on its grizzly bear population."[59]

In 2018, the Ninth Circuit similarly rejected FWS's decision not to list the Upper Missouri River Valley distinct population segment of Arctic grayling (*Thymallus arcticus*) as endangered or threatened.[60] In *Center for Biological Diversity v. Zinke*, the court held that FWS acted arbitrarily and capriciously when it failed to explain why the uncertainty of climate change effects on the Arctic grayling supported not listing it.[61] The court faulted FWS for (1) refusing to make any projections with respect to the synergistic effects of climate change "simply because of uncertainty," and (2) disregarding the additive effects of climate change in considering the effects of low stream flows and high water temperatures on the species.[62]

Other courts have similarly faulted FWS for requiring a greater level of scientific certainty or evidence than the ESA requires with respect to climate change effects on a species in its listing determination. In *Defenders of Wildlife v. Jewell*, a federal district court in Montana held that FWS's 2014 decision to withdraw the proposed listing of the North American wolverine (*Gulogulo luscus*) as threatened was arbitrary and capricious.[63] In reversing its position on the proposed listing, FWS

[59] *Id.* at 1026. On remand, FWS again delisted the grizzly bear population from the threatened species list in 2017. Removing the Greater Yellowstone Ecosystem Population of Grizzly Bears from the Federal List of Endangered & Threatened Wildlife, Final Rule, 82 Fed Reg. 30,502, 30,536-540 (June 30, 2017). The consideration of climate change played a lesser role in the district court's decision to vacate FWS's 2017 attempt to delist the Yellowstone grizzly bears from the threatened species list. *See* Crow Indian Tribe v. United States, 343 F. Supp. 3d 999, 1010-15 (D. Mont. 2018) (concluding that FWS failed to consider how reduced protections of the Yellowstone grizzly would affect other grizzlies outside the Greater Yellowstone Ecosystem). In agreeing with the plaintiff's claim that FWS failed to analyze the Yellowstone grizzlies' genetic health properly, the court noted that studies expressed concerns about the long-term viability of isolated grizzly populations and recommended cross-breeding between ecosystems "particularly given the unpredictability of future climate and habitat changes." *Id.* at 1020.

[60] Ctr. for Biological Diversity v. Zinke, 900 F.3d 1053, 1073 (9th Cir. 2018) ("By failing to explain why the uncertainty of climate change favors not listing the arctic grayling when the 2014 Finding acknowledges the warming of water temperatures and decreasing water flow because of global warming, FWS acted in an arbitrary and capricious manner.")

[61] *Id.* at 1072.

[62] *Id.* The Ninth Circuit remanded the determination to FWS for reassessment of its findings. *Id.* at 1073.

[63] Defs. of Wildlife v. Jewell, 176 F.Supp.3d 975, 1011 (D. Montana 2016). According to FWS, warming temperatures and declines in snow levels predicted by climate models are expected

attempted to discredit certain scientific studies on climate change effects that it had relied on previously to propose listing the wolverine as a threatened species.[64] FWS claimed that FWS needed greater certainty and refinement in the climate change data before listing the wolverine.[65] The court concluded that FWS "cannot demand a greater level of scientific certainty than has been achieved in the field to date—the 'best scientific data available' . . . standard does not require that the [FWS] act only when it can justify its decision with absolute confidence, and 'the ESA accepts agency decisions in the face of uncertainty.'"[66] After the court decision, in 2016, FWS reopened the public comment period on its previous proposal to list the wolverine as threatened under the ESA.[67] FWS has not made a final listing determination after closing the comment period.

Relatedly, courts have faulted FWS for requiring more evidence of climate change effects to delist a species than what is required under the ESA. In 2019, a federal district court in Texas held that FWS had acted arbitrarily and capriciously when it denied a petition to delist the Bone Cave harvestman spider (*Texella reyesi*) as an endangered species.[68] In *American Stewards of Liberty v. Department of the Interior*, the court concluded that FWS did not deny the petition based on the best available data but instead based its denial on the absence of "admittedly unavailable" evidence of climate change effects on the species and its habitat.[69] The court did not defer to FWS's conclusion that delisting of the spider was not warranted because the petition failed, in part, to include a "trend analysis to indicate that this species can withstand the threats associated with development or climate change over the long term."[70] In its decision

to affect the North American wolverine, which relies on deep snow to den and reproduce, over the coming decades. *Id.* at 978-80.

[64] *Id.* at 1002-03.

[65] *Id.*

[66] *Id.* at 1003.

[67] Endangered and Threatened Wildlife and Plants; Proposed Rule for the North American Wolverine; Proposed rule; reopening of comment period, 81 Fed. Reg. 71,670 (Oct. 16, 2016).

[68] Am. Stewards of Liberty v. Dept. of the Interior, 370 F. Supp. 3d 711, 728-730 (W.D. Tex. 2019).

[69] *Id.* at 727.

[70] *Id.*

denying the petition to delist, FWS claimed that the petitioners did not present enough data to determine if the spider's population will continue to decline from such threats.[71] The court held that FWS "committed clear error" by requiring the petition to present "conclusive evidence about the harvestman's population trends—more evidence than the Service admits is available or attainable."[72] The court concluded that the petition met the threshold for a finding that delisting may be warranted and remanded to FWS for further consideration.[73] To date, FWS has not issued a new finding regarding the delisting petition.

Although it is not possible to have complete certainty of future climate change effects, these court decisions illustrate that the Services cannot rely solely on scientific uncertainty to make listing or delisting decisions without adequate justification. In the 2019 revised ESA regulations, the Services noted that "the requirement to use the 'best available' data means that we cannot insist that information must be free from all uncertainty, and further agree that the Act's protections should not be withheld until a species' status has declined to the point that the future risk of extinction is certain."[74]

DESIGNATING CRITICAL HABITAT

The Services have also considered climate change effects in designating critical habitat. When listing a species as threatened or endangered, the ESA requires the Services to "designate any habitat of such species which is then considered to be critical habitat."[75] As a threshold matter, as made clear by the Supreme Court's 2018 decision in *Weyerhauser Co. v. FWS*, an area must be "habitat" for a species for the

---

[71] *Id.*

[72] *Id.*

[73] *Id.* at 735.

[74] Endangered and Threatened Wildlife and Plants; Regulations for Listing Species and Designating Critical Habitat, Final Rule, 84 Fed. Reg. 45,020, 45,033 (Aug. 27, 2019).

[75] 16 U.S.C. §1533(a)(3)(A)(i).

Services to consider whether it is "critical habitat."[76] Under the ESA, the Services may designate two types of habitat as critical habitat: (1) specific areas within the geographical area occupied by the species, which contain the "physical or biological features essential to the conservation of the species" and may require special management protections (occupied habitat); and (2) areas outside the geographical areas occupied by the species if the Secretary determines that such unoccupied areas are "essential for the conservation of the species" (unoccupied habitat).[77] Once an area is designated as critical habitat, federal agencies may not (unless exempted) authorize, fund, or carry out actions that are likely to "result in the destruction or adverse modification" of critical habitat.[78]

The Services face unique challenges when designating critical habitat based on modeled habitat shifts for species affected by climate change.[79] The legal challenges to FWS's designation of the polar bear's critical habitat show how a court deferred to the FWS's interpretation of climate change data and models to determine whether unoccupied areas are "essential for the conservation of the species."

In a 2016 decision, *Alaska Oil & Gas Association v. Jewell*, the Ninth Circuit upheld FWS's designation of 187,000 square miles as critical habitat for the polar bear.[80] FWS based its critical habitat designation in part on long-term projections of habitat destruction from climate change.[81] FWS designated three areas on Alaska's coast and in its waters that contain elements essential to the polar bear: a sea ice habitat, a terrestrial denning

---

[76] Weyerhaeuser Co. v. FWS, 139 S. Ct. 361, 368 (2018). While acknowledging that the ESA does not provide a "baseline definition of habitat," the Supreme Court left the task to the lower court to interpret the term "habitat." *Id.* at 369.

[77] 16 U.S.C. §1532(5)(A)(i). The ESA defines conservation as the "use of all methods and procedures which are necessary to bring any endangered species or threatened species to the point at which the measures provided pursuant to this chapter are no longer necessary." *Id.* § 1532(3).

[78] *Id.* § 1536(a)(2).

[79] *See* David Owen, *Endangered Species Act, in* GLOBAL CLIMATE CHANGE AND U.S. LAW 190-91 (Michael B. Gerrard & Jody Freeman eds., 2d ed. 2014) (discussing the scope of habitat designations and the difficulty in predicting habitat shifts); LAWRENCE R. LIEBESMAN & RAFE PETERSEN, ENDANGERED SPECIES DESKBOOK 106 (2010) (explaining the use of predictive modeling to identify habitat shifts).

[80] Alaska Oil & Gas Ass'n v. Jewell, 815 F.3d 544, 558-59 (9th Cir. 2016).

[81] *Id.* at 558-59.

habitat, and a barrier island habitat.[82] For two of the designated areas, the district court concluded that FWS failed to provide evidence that the two areas included all of the elements required for the survival of the polar bear.[83] The district court asked FWS to establish that polar bears currently use those two areas as habitat.[84]

The Ninth Circuit disagreed with the lower court's narrow interpretation of the ESA critical habitat requirements. The court rejected the lower court's finding that the ESA required FWS to limit the critical habitat designation to specific areas that are currently used by polar bears, explaining that "[n]o such limitation to existing use appears in the ESA, and such a narrow construction of critical habitat runs directly counter to the Act's conservation purposes. The Act is concerned with protecting the future of the species, not merely the preservation of existing bears. And it requires use of the best available technology, not perfection."[85] The court concluded that FWS properly relied on climate science and sea ice data in designating habitat that has the elements required to sustain and preserve the polar bear population.[86]

Similar to cases regarding foreseeability and scientific uncertainty, the court appeared to defer to FWS's reasoned consideration of climate change effects based on evidence in the record.

Revised ESA Critical Habitat Regulations

The 2019 final rule clarified when the Secretary may designate unoccupied areas as critical habitat. Under the ESA, unoccupied areas must be essential to the conservation of the species to be critical habitat.[87] Under

---

[82] *Id.* at 555.

[83] Alaska Oil & Gas Ass'n v. Salazar, 916 F. Supp. 2d 974, 1001 (D. Alaska 2013).

[84] *Id.*

[85] *Id.* at 555-56.

[86] *See id.* at 558-59 (listing the studies, reports, and climate models that predict sea ice loss and rescission as a result of climate change). *Cf.* All. for the Wild Rockies v. Lyder, 728 F. Supp. 2d 1126, 1139-40 (D. Mont. 2010) (upholding FWS's decision not to designate unoccupied Canada lynx habitat to account for climate change because the science did not provide the specificity needed to identify the location of lynx habitat in the future).

[87] 16 U.S.C. §1532(5)(A)(i).

the revised regulations, to determine if an unoccupied area is essential, the Services must find that the occupied habitat of the species at the time of listing is inadequate to ensure the conservation of the species.[88] The Services must also determine that there is a "reasonable certainty" that the area will (1) contribute to the conservation of the species, and (2) contains one or more of those physical or biological features essential to the conservation of the species.[89] The final rule explains that this revision "better reflects the need for high confidence that an area designated as unoccupied critical habitat will actually contribute to the conservation of the species."[90]

How the revised regulations will affect the designation of unoccupied critical habitat will likely depend on the threshold the Services set for "reasonable certainty" that the unoccupied habitat will contribute to the conservation of the species. Some stakeholders are concerned that these changes to the critical habitat regulations may limit the Services' ability to protect species that move because of climate-change-related habitat loss.[91] In the litigation challenging the 2019 final rules, the plaintiffs argue that, by imposing an "elevated certainty requirement" on the Services' determination of what areas are "essential," the new rules would preclude the Services from designating currently unoccupied areas to which species may need to move because of climate change as critical habitat.[92] In contrast, other stakeholders see the regulatory changes as complying with the Supreme Court's decision in *Weyerhauser Co. v. FWS* that an area must be "habitat" before the Services may consider whether it is "critical habitat."[93] These stakeholders assert that reasonable certainty that an area has at least one of the essential features necessary to conserve the species

---

[88] Endangered and Threatened Wildlife and Plants; Regulations for Listing Species and Designating Critical Habitat, Final Rule, 84 Fed. Reg. 45,020, 45,043-45 (Aug. 27, 2019).

[89] *Id.* at 45,043.

[90] *Id.* at 45,045.

[91] *Id.* at 45,043.

[92] Complaint for Declaratory and Injunctive Relief at 20-21, Center for Biological Diversity v. Bernhardt, No. 3:19-cv- 05206 (N.D. Cal. Aug. 21, 2019).

[93] Bonner R. Cohen, *Endangered Species Act Reforms Will Benefit Wildlife and People*, THE HILL (Aug. 16, 2019), https://thehill.com/opinion/energy-environment/457534-endangered-species-act-reforms-will-benefit-wildlife-and- people.

ensures that the area is habitat for the species.[94] In addition, landowners claim that these changes remove potential regulatory burdens that critical habitat designations cause, such as requirements that, when issuing permits that may adversely affect critical habitat, federal agencies consult with stakeholders.[95]

PROTECTING ENDANGERED OR THREATENED SPECIES

If FWS or NMFS bases its listing decision on climate change effects, FWS or NMFS must also determine whether federal actions that contribute to climate change jeopardize the species under ESA Section 7 or whether an entity that may contribute to climate change is "taking" the species in violation of ESA Section 9. The Services may tailor the Section 9 "take" prohibitions for species listed as threatened under the ESA by using Section 4(d) rules. This section reviews how the Services address climate change effects when protecting listed species under ESA Sections 4(d), 7, and 9.

Prohibiting "Take" Under Sections 9 and 4(d)

ESA Section 9 prohibitions on "taking" a listed species differ for threatened and endangered species. ESA Section 9(a)(1) prohibits the unauthorized "take" of an *endangered* species. *Take* is defined as an act "to harass, harm, pursue, hunt, shoot, wound, kill, trap, capture, or collect or to attempt to engage in any such conduct."[96] In contrast, the ESA does not prohibit the taking of a *threatened* species unless FWS or NMFS decides to extend the Section 9 take prohibitions to the threatened species through a

---

[94] *Id. See also* Press Release, Dep't of the Interior, What They Are Saying – Endangered Species Act Announcement, https://www.doi.gov/pressreleases/what-they-are-saying-endangered-species-act-announcement (last visited Sept. 19, 2019).

[95] *Id.*

[96] 16 U.S.C. § 1538(a)(1); 1532(19).

Section 4(d) rule.[97] For threatened species, ESA Section 4(d) requires FWS or NMFS to issue regulations it "deems necessary and advisable to provide for the conservation of such species."[98] In 1978, FWS issued a "blanket 4(d) rule" that extends most of the Section 9 take prohibitions to all threatened species listed by FWS, unless it adopts a specific rule for a particular species.[99] As discussed below, the 2019 revisions to the ESA regulations rescinded FWS's blanket 4(d) rule for newly listed or reclassified species, aligning it with NMFS's practice of issuing species-specific 4(d) rules for threatened species.[100]

Climate Change and Section 4(d) Rules for Threatened Species

A frequent debate among legal scholars and stakeholders is whether the take prohibitions should extend to GHG-emitting activities that contribute to climate change.[101] Stakeholders seeking greater protection for species argue that sources of GHG emissions cause an unlawful "take" under ESA Section 9 because GHG emissions contribute to climate change, which harms the species.[102] However, the Services and critics of this approach assert that the Section 9 take prohibitions can apply only if GHG-emitting activity directly and intentionally takes the species or negatively affects its habitat.[103]

In the litigation challenging the polar bear's 4(d) rule, the federal court's decision highlighted the challenges in applying the take prohibitions to GHG-emitting activities. Plaintiffs challenged FWS's 4(d) rule that specified prohibitions necessary to conserve the threatened polar

---

[97] *Id.* § 1533(d).

[98] *Id.* § 1533(d).

[99] 50 C.F.R. § 17.31(a), (c); Protection for Threatened Species of Wildlife; Final Rule, 43 Fed. Reg. 18,180 (April 28, 1978).

[100] Endangered and Threatened Wildlife and Plants; Regulations for Prohibitions to Threatened Wildlife and Plants, Final Rule, 84 Fed. Reg. 44,753, 44,753-54 (Aug. 27, 2019).

[101] Michael C. Blumm & Kya B. Marienfeld, *Endangered Species Act Listings and Climate Change: Avoiding the Elephant in the Room*, 20 ANIMAL L. 277, 298-300 (2014).

[102] *See* David Owen, *Endangered Species Act, in* GLOBAL CLIMATE CHANGE AND U.S. LAW 195 (Michael B. Gerrard & Jody Freeman eds., 2d ed. 2014).

[103] *Id.*

bear species.[104] The rule, among other things, did not prohibit activities outside the species' current range that may incidentally affect polar bears, such as GHG-emitting activities that may contribute to the loss of sea ice habitat.[105] Plaintiffs claimed that the rule was arbitrary and capricious and violated the ESA by failing to address threats to the polar bear from GHG emissions and the loss of potential sea ice habitat outside the polar bears' range.[106]

In rejecting this argument, the court concluded FWS had a rational basis not to extend the ESA's take prohibitions because there was insufficient evidence to suggest that regulating offsite GHG- producing activities would produce direct conservation benefits to the polar bear.[107] FWS explained that the best available science and climate modeling could not identify an individual GHG emission source as the cause of a specific adverse effect on the polar bear or its habitat.[108] The court acknowledged that it cannot "decide based upon its own independent assessment" "whether the ESA is an effective or appropriate tool to address the threat of climate change . . . . The answer to that question will ultimately be grounded in science and policy determinations that are beyond the purview of this Court."[109]

Based on this judicial opinion, it seems unlikely that the Services will use Section 4(d) rules to prohibit GHG-emitting activities without further advances in science that can establish a causal connection between the

---

[104] *In re* Polar Bear Endangered Species Act Listing & § 4(d) Rule Litig., 818 F. Supp. 2d. 214, 218 (D.D.C. 2011); Special Rule for the Polar Bear, 73 Fed. Reg. 76,249 (Dec. 16, 2008).

[105] 73 Fed. Reg. at 76251.

[106] *In re* Polar Bear Endangered Species Act Listing & § 4(d) Rule Litig., 818 F. Supp. 2d. 214, 218 (D.D.C. 2011) (highlighting plaintiffs' contention that FWS "[could not] effectively provide for the conservation of the polar bear without addressing global greenhouse gas emissions, which the agency itself identified as the cause of increasing Arctic temperatures that are expected to lead to a significant decline of the polar bear's sea ice habitat").

[107] *Id.* at 231-33 ("The Service found no evidence to suggest that extending the ESA incidental take provisions outside the range of the polar bear would produce similar conservation benefits, however. With respect to these indirect impacts, in the event that an incidental take can be identified and attributed to a specific cause originating outside the species' range, the Service found that the incidental take provisions of the MMPA are sufficient to address that violation.").

[108] 73 Fed. Reg. at 76266.

[109] *In re* Polar Bear Endangered Species Act Listing, 818 F. Supp. 2d at 234.

individual GHG emission source and the specific adverse effect on the species or its habitat.

Rescinding FWS's Blanket 4(d) Rule

In 2019, FWS rescinded the "blanket 4(d) rule" for newly listed or reclassified threatened species, and will now adopt species-specific 4(d) rules.[110] Because the rescission applies prospectively, the blanket 4(d) rule continues to prohibit the take of threatened species covered by the blanket 4(d) rule that FWS listed prior to the effective date of the rescission.[111] This species-specific approach aligns with NMFS's practice of establishing specific 4(d) rules for each threatened species.[112]

How the rescission of FWS's blanket 4(d) rule may affect the protection of species threatened by climate change effects depends on its implementation. While some stakeholders are concerned that the rescission will "weaken" protections for threatened species because of delays in issuing species-specific 4(d) rules, it may have little effect on whether GHG-emitting activities are prohibited.[113] FWS has not adopted a 4(d) rule that prohibited GHG-emitting activities that could affect threatened species and their habitats, prohibiting only actions that directly and intentionally take threatened species.[114] For threatened species affected by climate change, legal scholars argue that such "limited" 4(d) rules have "no real

---

[110] Endangered and Threatened Wildlife and Plants; Regulations for Prohibitions to Threatened Wildlife and Plants, Final Rule, 84 Fed. Reg. 44,753, 44,753-54 (Aug. 27, 2019).

[111] *Id.*

[112] *Id.*

[113] *See* Complaint for Declaratory and Injunctive Relief at 3, Center for Biological Diversity v. Bernhardt, No. 3:19-cv- 05206 (N.D. Cal. Aug. 21, 2019). *See also* Jake Li, *Last Week's Endangered Species Regulations: What Really Happened?* ENVTL. POLICY & INNOVATION CENTER (Aug. 20, 2019), http://policyinnovation.org/the-real-story-on-the- new-endangered-species-act-rules/; Jonathan Wood, The New Endangered Species Act Rules, Explained, PROPERTY AND ENV'T RESEARCH CENTER (Aug. 14, 2019), https://www.perc.org/2019/08/14/the-new-endangered-species-act- rules-explained/.

[114] *See* Michael C. Blumm & Kya B. Marienfeld, *Endangered Species Act Listings and Climate Change: Avoiding the Elephant in the Room*, 20 ANIMAL L. 277, 298-299 (2014) (discussing the lack of taking prohibitions related to GHG emissions in the polar bear 4(d) rule and the proposed wolverine 4(d) rule).

effect on the activities that are causing climate change, the acknowledged primary factor contributing to [the] species' decline."[115]

Section 7 Consultation

Some stakeholders and legal scholars view the ESA Section 7 consultation requirement as a potentially powerful tool to limit GHG-emitting activities that may further jeopardize threatened or endangered species that were listed, at least in part, because of climate change effects.[116] In practice, the Services and the courts have acknowledged that climate change should be considered during the consultation process.[117] However, the courts have not required the Services to curb activities that may contribute to climate change to protect threatened or endangered species.

In general, ESA Section 7 requires federal agencies to "insure that any action authorized, funded, or carried out by such agency . . . is not likely to jeopardize the continued existence of any endangered species or threatened species or result in the destruction or adverse modification of [the critical] habitat of such species."[118] A federal agency planning any action must consult with NMFS or FWS if the federal agency determines that its action "may" jeopardize a listed species or adversely affect its habitat.[119] The ESA and its implementing regulations specify the types of consultation (e.g., informal versus formal consultation), when each type of consultation is required, and the procedures the agency proposing the action and the Services must follow.[120]

---

[115] *Id.*

[116] *See, e.g.*, James Ming Chen, *The Fragile Menagerie: Biodiversity Loss, Climate Change, and the Law*, 93 IND. L.J. 303, 334 -35 (2018).

[117] *See* David Owen, *Endangered Species Act, in* GLOBAL CLIMATE CHANGE AND U.S. LAW 193-94 (Michael B. Gerrard & Jody Freeman eds., 2d ed. 2014) (concluding that Section 7 has played "no role" in mitigating GHG emissions).

[118] 16 U.S.C. § 1536(a)(2).

[119] *Id.* § 1536(a)(2); 50 C.F.R. §402.14(a).

[120] For additional information on Section 7 consultation procedures, see CRS In Focus IF11241, *The Legal Framework of the Endangered Species Act (ESA)*, by Erin H. Ward; and CRS Report RL31654, *The Endangered Species Act: A Primer*, by Pervaze A. Sheikh.

After the consultation with an agency, the Services must issue a biological opinion (BiOp) based on "the best scientific and commercial data available" that determines whether the proposed action is likely to jeopardize the ESA-listed species or adversely modify critical habitat.[121] If the Services determine that an agency action would likely jeopardize the listed species or its critical habitat, the agency must terminate the action, implement a Service-proposed alternative action, or seek an exemption.[122] If the agency action is not likely to jeopardize the continued existence of the species but is nonetheless likely to result in some "incidental take" of the species, the BiOp must set forth an incidental take statement, which specifies the permissible "amount or extent" of this effect on the species.[123]

Various court decisions have faulted the Services for failing to discuss climate change effects when assessing whether federal action will jeopardize a listed species or adversely modify its habitat.[124] In the 2007 decision, *Natural Resources Defense Council v. Kempthorne*, a federal district court in California held that FWS acted arbitrarily and capriciously when analyzing potential effects on the threatened delta smelt *(Hypomesus transpacificus)* from a large water diversion project.[125] The court determined that the "absence of <u>any</u> discussion in the BiOp of how to deal with any climate change is a failure to analyze a potentially 'important aspect of the problem.'"[126] In rejecting FWS's claim that the climate change studies did not merit analysis because they were inconclusive, the court concluded that without any meaningful discussion in the BiOp, it was "impossible" for the court to determine whether the climate studies were "rationally discounted because of [their] inconclusive nature, or arbitrarily ignored."[127]

---

[121] 50 C.F.R. § 402.14(g)(8), (h)(2)–(3).

[122] *Id.* § 402.15(a)-(c).

[123] *Id.* § 402.14(i).

[124] *Id.* § 402.14(g); Pac. Coast Fed'n of Fishermen's Ass'ns v. Gutierrez, 606 F. Supp. 2d 1122, 1184 (E.D. Cal. 2008); Nat. Res. Def. Council v. Kempthorne, 506 F. Supp. 2 322, 374-76 (E.D. Cal. 2007).

[125] *Kempthorne*, 506 F. Supp. 2d at 370.

[126] *Id.* (quoting Pac. Coast Fed'n of Fishermen's Ass'ns v. Gutierrez, 606 F. Supp. 2d 1122, 1134 (E.D. Cal. 2008)) (emphasis in original).

[127] *Id.* at 369.

Similarly, plaintiffs successfully challenged NMFS's BiOp that concluded that changes to a fish hatchery operation were not likely to jeopardize the species or adversely affect critical habitat for the Upper Columbia River spring Chinook salmon (*Oncorynchus tshawytscha*) or steelhead (*Oncorhynchus mykiss*).[128] A federal district court in Washington ruled that NMFS's BiOp was arbitrary and capricious because it failed to analyze adequately climate change effects from the hatchery's modified operations and water use.[129] The court explained that "[t]he best available science indicates that climate change will affect stream flow and water conditions throughout the Northwest" and that the lack of a model or study specifically addressing local climate change effects did not permit NMFS to ignore this factor.[130] The court found that NMFS had included "no discussion whatsoever" of the potential effects of climate change on the hatchery's future operations and water use, and that it was not sufficient for NMFS to say that the local area at issue was less prone to climate change effects than other areas in the region.[131]

When the Services have discussed climate change effects from federal actions, some courts have scrutinized the Services' rationale in dismissing such effects when issuing a "no jeopardy" BiOp. For example, the Ninth Circuit 2017 majority opinion in *Turtle Island Restoration Network v.*

Department of Commerce[132] examined NMFS's BiOp that concluded that a fishery expansion would neither jeopardize the continued existence of the endangered loggerhead sea turtle (*Caretta caretta*) nor the endangered leatherback sea turtle (*Dermochelys coriacea*). For the loggerhead sea turtles, the court ruled that NMFS had acted arbitrarily and capriciously by failing to incorporate into its jeopardy analysis climate-model data that predicted that the fishery expansion would "exacerbate" loggerhead population decline due to climate change.[133] In contrast, for the leatherback sea turtles, the majority upheld NMFS's no-jeopardy

---

128 Wild Fish Conservancy v. Irving, 221 F. Supp. 3d 1224 (E.D. Wash. 2016).
129 *Id.* at 1228.
130 *Id.* at 1233.
131 *Id.* at 1233-34. The judge remanded the matter for further consultation. *Id.* at 1234.
132 Turtle Island Restoration Network v. Dep't of Commerce, 878 F.3d 725 (9th Cir. 2017).
133 *Id.* at 738-39.

conclusion, rejecting the plaintiffs' argument that NMFS erred by limiting the "temporal scale" of its analysis to 25 years despite NMFS's determination that rising temperatures from climate change would have effects on leatherback sea turtles over the next century.[134] Because NMFS's BiOp considered and concluded that it could not credibly predict climate change effects on the leatherback turtles, the majority held that NMFS adequately considered the climate change effects in its no-jeopardy conclusion.[135]

Despite some success challenging BiOps, neither the courts nor the Services have found that climate change effects from a proposed federal action jeopardize the species or adversely modify its habitat.[136] Some stakeholders and legal scholars argue that when a proposed federal action contributes to climate change that may jeopardize the species or adversely modify its habitat, the agency is required to consult with the Services.[137] If the Services determine that such actions jeopardize the species or its habitat, these stakeholders assert that the Services should use Section 7 consultation authority to limit or modify the GHG emissions from the proposed federal action.[138] However, the Department of the Interior issued a Solicitor's Opinion explaining that Section 7 consultation is not required if no causal connection exists among the proposed federal action, a reasonably certain climate change effect, and the listed species.[139] Therefore, without evidence of a causal connection between the proposed action and climate change effects, Section 7 consultation will not be triggered, foreclosing any opportunity for the Services to consider

---

[134] *Id.* at 739.

[135] *Id.* at 740-41.

[136] David Owen, *Endangered Species Act, in* GLOBAL CLIMATE CHANGE AND U.S. LAW 194 (Michael B. Gerrard & Jody Freeman eds., 2d ed. 2014).

[137] *See* James Ming Chen, *The Fragile Menagerie: Biodiversity Loss, Climate Change, and the Law*, 93 IND. L.J. 303, 335 (2018) (discussing how greater enforcement of Section 7 consultation requirements could have a greater effect on climate mitigation); David Owen, *Endangered Species Act, in* GLOBAL CLIMATE CHANGE AND U.S. LAW 192 (Michael B. Gerrard & Jody Freeman eds., 2d ed. 2014) (analyzing stakeholder arguments for using Section 7 to address GHG- emitting activities).

[138] *Id.*

[139] *See* Memorandum from David Longly Bernhardt, Solicitor, Dept. of the Interior, *Guidance on the Applicability of the Endangered Species Act's Consultation Requirements to Proposed Actions Involving the Emission of Greenhouse Gases* (Oct. 3, 2008).

mitigating the climate change effects from such actions. Federal agencies and the Services have continued to use this policy to comply with their Section 7 consultation obligations.[140]

Revising Section 7 Consultation Regulations

The 2019 ESA regulation revisions codified the Services' existing Section 7 climate change policy. Existing ESA Section 7 regulations require the federal agency proposing the action and the Services to evaluate the status of the listed species or critical habitat, the "effects of the action," and cumulative effects.[141] Prior to the 2019 revisions, ESA regulations defined "effects of the action" to include both direct and indirect effects of a proposed federal action on the species or critical habitat.[142] The 2019 ESA rule revised the definition of "effects of the action" to include all consequences to listed species or critical habitat that are *caused by* the proposed action.[143] The definition specified that a consequence is "caused by the proposed action if it would not occur but for the proposed action and it is reasonably certain to occur."[144] The Services provided a two-part test to identify a consequence: (1) whether the effect or activity would not occur but for the action and (2) whether the effect or activity is reasonably certain to result from the action.[145] The preamble explains that "if the agency fails to take the proposed action, and the activity would still occur, there is not 'but for' causation."[146]

Some stakeholders support the revisions to the Section 7 consultation requirements, asserting that the changes will "help decrease the resources

---

[140] *See, e.g.*, Greenhouse Gas Emissions and Fuel Efficiency Standards for Medium- and Heavy-Duty Engines and Vehicles—Phase 2, Final Rule, 81 Fed. Reg. 73,478, 73,969 (Oct. 25, 2016) (concluding that the best scientific data available were insufficient to draw a causal connection between GHG emissions and effects on the species in its habitat).

[141] 50 C.F.R. § 402.14(g)(2)-(3).

[142] *Id.*

[143] Endangered and Threatened Wildlife and Plants; Regulations for Interagency Cooperation, Final Rule, 84 Fed. Reg. 44,976, 44,976-77 (Aug. 27, 2019).

[144] *Id.* at 45,016 (to be codified in 50 C.F.R. § 402.02).

[145] *Id.* at 44,978.

[146] *Id.* at 44,977.

needed for federal agencies and applicants to describe the effects of their actions to listed species or critical habitat when engaged in section 7 consultation."[147] Other stakeholders contend that the proposed changes will "unreasonably narrow" and "bar" Section 7 consultation when climate change effects do not affect immediately the geographic area of the project.[148]

POTENTIAL IMPLICATIONS

When the ESA was enacted in 1973, Congress did not consider climate change as a significant factor in conserving endangered species.[149] Although the Services and the courts have acknowledged that actions taken under the ESA must consider climate change effects on species and their habitats, the debate continues on whether the ESA can adequately protect and conserve species threatened by climate change effects. Stakeholders disagree on what role the ESA should play in addressing climate change, with some arguing that the ESA is not equipped to mitigate climate change effects.[150] Other stakeholders believe that the Services can and should wield the ESA to protect species threatened by climate change and to curb activities contributing to climate change.[151]

---

[147] Press Release, Sen. Ted Cruz, Cruz, Colleagues Praise Trump Administration's Act to Modernize Broken ESA Processes Negatively Impacting Species and Landowners (Aug. 12, 2019), https://www.cruz.senate.gov/?p=press_release&id=4626.

[148] Press Release, Def. of Wildlife, Trump Administration Releases Regulations Endangering Imperiled Wildlife (Aug. 12, 2019), https://defenders.org/newsroom/trump-administration-releases-regulations-endangering-imperiled-wildlife; Complaint for Declaratory and Injunctive Relief at 23-28, Center for Biological Diversity v. Bernhardt, No. 3:19-cv- 05206 (N.D. Cal. Aug. 21, 2019).

[149] Barry Kellman, *Climate Change in the Endangered Species Act: A Jurisprudential Enigma*, 46 ENVTL. L. REP. 10845, 10854 (2016); David Owen, *Endangered Species Act, in* GLOBAL CLIMATE CHANGE AND U.S. LAW 196 (Michael B. Gerrard & Jody Freeman eds., 2d ed. 2014).

[150] *See* David Owen, *Endangered Species Act, in* GLOBAL CLIMATE CHANGE AND U.S. LAW 188-196 (Michael B. Gerrard & Jody Freeman eds., 2d ed. 2014) (discussing competing views on the effectiveness of using ESA to address climate change).

[151] *Id. See also* J.B. Ruhl, Climate Change and the Endangered Species Act: Building Bridges to the No-Analog Future, 39 ENTVL. L. REP. 10,735, 10,739-44 (discussing policy drivers to integrate climate change considerations and actions into the ESA).

Generally, legal scholars agree that litigation has influenced how the Services factor climate change effects into ESA decisions.[152] Legal challenges have helped to ensure that the Services consider projected climate change effects on species in their ESA decisions. In light of the judicial deference afforded to the Services, the courts have not expanded the ESA as a tool to protect listed species by regulating activities that contribute to climate change.[153] From the Services' viewpoint, the best available scientific and commercial data have been insufficient to determine that GHG emissions from a proposed activity cause detrimental effects on the species or its habitat.[154] However, as climate modeling and technology advance, the Services may be able to predict the causes and effects from climate change on species with greater scientific certainty and data.

Members of Congress may be interested in the implications of revising the ESA to clarify its treatment of climate change effects.[155] Legislation could clarify whether ESA Section 9 prohibitions or Section 7 consultation requirements apply to indirect harms that contribute to climate changes that may affect a species' survival, or how the Services should address scientific uncertainty associated with projected climate change effects when making listing determinations. As legislative proposals continue to develop, legal battles over the how the Services interpret climate change effects in their ESA decisions will likely continue.

---

[152] Barry Kellman, *Climate Change in the Endangered Species Act: A Jurisprudential Enigma*, 46 ENVTL. L. REP. 10845, 10854 (2016); *See also* Jennifer Hijazi, *Climate Change Looms Large in Endangered Species Litigation*, E&E NEWS (July 2, 2019) (describing various legal petitions challenge the Services' ESA decisions regarding climate change), https://www.eenews.net/stories/1060682935.

[153] David Owen, *Endangered Species Act, in* GLOBAL CLIMATE CHANGE AND U.S. LAW 196 (Michael B. Gerrard & Jody Freeman eds., 2d ed. 2014)

[154] Endangered and Threatened Wildlife and Plants; Regulations for Interagency Cooperation, Final Rule, 84 Fed. Reg. 44,976, 46,991 (Aug. 27, 2019).

[155] *See* Kellie Lunney, *Dems, GOP will Duke It Out on ESA Overhaul*, E&E DAILY (Sept. 18, 2019), https://www.eenews.net/eedaily/2019/09/18/stories/1061132049 (discussing various legislative proposals seeking to revise the ESA).

In: Climate Change
Editor: Jeremy Nesbitt
ISBN: 978-1-53617-216-4

Chapter 6

THE GREEN CLIMATE FUND (GCF)*

Richard K. Lattanzio

INTERNATIONAL ENVIRONMENTAL ASSISTANCE

Many governments hold that environmental degradation and climate change pose international and trans-boundary risks to human populations, economies, and ecosystems. To confront these challenges, governments have negotiated various international agreements to protect the environment, reduce pollution, conserve natural resources, and promote sustainable growth. While some observers call upon industrialized countries to take the lead in addressing these issues, many recognize that efforts are unlikely to be sufficient without similar measures being taken in lower- income countries. However, lower-income countries, which tend to focus on poverty reduction and economic growth, may not have the financial resources, technological know- how, and/or institutional capacity to deploy environmentally protective measures on their own.

* This is an edited, reformatted and augmented version of Congressional Research Service Publication No. IF10382, updated October 2, 2019.

Therefore, international financial assistance, or foreign aid, has been a principal method for governments to support actions on global environmental problems in lower-income countries. Often, this assistance can serve as a cost-effective strategy for donor countries to provide greater market access for domestic goods and services abroad and increased environmental benefits at home.

The United States and other industrialized countries have committed to providing financial assistance for global environmental initiatives through a variety of multilateral agreements, including the Montreal Protocol (1987), the U.N. Framework Convention on Climate Change (1992), and the U.N. Convention to Combat Desertification (1994). International financial assistance takes many forms, from fiscal transfers to market transactions. It may include grants, loans, loan guarantees, export credits, insurance products, and private sector investment. It may be structured as official bilateral development assistance or as contributions to multilateral development banks and other international financial institutions.

THE UNITED NATIONS FRAMEWORK CONVENTION ON CLIMATE CHANGE (UNFCCC)

The 1992 UNFCCC was the first formal international agreement to acknowledge and address human-driven climate change. The U.S. Senate provided its advice and consent to the convention's ratification in 1992, the same year it was concluded (Treaty Number: 102-38, October 7, 1992). The UNFCCC entered into force in 1994. As of March 2016, 196 governments are parties to the UNFCCC.

As a framework convention, the UNFCCC provides general obligations for all parties but leaves them free to design their own means to fulfill these obligations. The convention "acknowledges" that climate change is a "common concern to humankind" and, accordingly, requires parties to (1) inventory, report, and control their human-related greenhouse gas emissions; (2) cooperate in preparing to adapt to climate change; and (3) assess and review the effective implementation of the convention. The

UNFCCC also commits the higher-income parties (i.e., those listed in Annex II of the convention as members of the Organization for Economic Cooperation and Development in 1992) to seek to mobilize financial assistance to help lower-income countries meet certain obligations common to all parties.

THE GREEN CLIMATE FUND (GCF)

The GCF is one of the international financial institutions connected to the UNFCCC. The fund was proposed by parties to the UNFCCC during the 2009 Conference of Parties (COP) in Copenhagen, Denmark, and its design was agreed to by all parties during the 2011 COP in Durban, South Africa. The GCF was made operational in 2014. The GCF aims to assist lower-income countries in their efforts to combat climate change through the provision of grants and other concessional financing for mitigation and adaptation projects, programs, policies, and activities. The GCF is capitalized by contributions from donor countries and other sources, potentially including innovative mechanisms and the private sector.

The GCF currently complements many of the existing multilateral climate change funds (e.g., the Global Environment Facility, the Climate Investment Funds, and the Adaptation Fund). However, some parties believe that, as one of the official financial mechanisms of the UNFCCC, it may eventually replace or subsume the other funds. Many countries, specifically lower-income countries, expect that the GCF will become very large (i.e., in the range of several tens of billions to over $100 billion annually) and serve as the predominant institution for climate change assistance in the developing world. These countries believe that the agreement to establish the GCF within the framework of the United Nations has been a key success in the recent international negotiations. Other countries, however, would rather see a smaller role for the GCF, preferring instead to have private investment and other financing flow through a variety of channels. Some parties also caution that ambitious steps need to be taken to ensure that the GCF is operated appropriately, meeting a strict set of fiduciary standards and social safeguards, in order to achieve an adequate buy-in by donor countries of its effectiveness and by recipient countries of its legitimacy.

Information about GCF activities, organization, policies, projects, and contributions is available on its website at http://www.greenclimate.fund/.

Organizational Structure

The structure of the GCF was negotiated at various COP meetings to the UNFCCC. Decisions included:

- Status. Parties to the UNFCCC designated the GCF as "the operating entity of the Financial Mechanism of the Convention" to be "accountable to and function under the guidance of the COP to support projects, programmes, policies and other activities in developing country Parties" (UNFCCC Decision 3/CP.17§3, 2011).
- Governance. The UNFCCC set forth the composition of a board to have 24 members, composed of an equal number from lower- and higher-income country parties, with representation from relevant U.N. groupings including Small Island Developing States and Least Developed Countries (3/CP.17§A9). Functions of the board are to include approving projects and funding, selecting implementing agencies, monitoring fiduciary standards and environmental and social safeguards, and evaluating performance.
- Host country. The UNFCCC selected Songdo, Incheon, Republic of Korea, as the host of the GCF (6/CP.18§3).
- Management. The UNFCCC established an independent secretariat to execute the day-to-day operations of the fund (4/CP.19§2).
- Trustee. The UNFCCC invited the World Bank to serve as the interim trustee (3/CP.17§16).

FUNDING

The GCF was officially opened for capitalization at the U.N. Climate Summit in September 2014. Parties called for an immediate capitalization of between $10 billion and $15 billion over the course of the first year. Initial funding came from Germany, France, and a dozen other countries that pledged approximately $2.3 billion. Further pledges brought the total to approximately $10 billion by the close of 2014, including (in approximate U.S. dollars) Japan ($1.5 billion), United Kingdom ($1.2 billion), Germany ($1 billion), France ($1 billion), Sweden ($600 million), Canada ($300 million), Norway ($250 million), and Australia ($200 million), among others. The GCF Board launched its first replenishment cycle in October 2018.

U.S. Commitments

The Obama Administration announced a U.S. pledge of $3 billion over four years during the G-20 meetings in Australia on November 15, 2014. The Obama Administration's FY2016 budget requested $500 million for the fund. H.R. 2029, the Continuing Appropriations Act, 2016, enacted December 18, 2015, as P.L. 114-113, had no provisions explicitly providing—or prohibiting—funding for the GCF. The Obama Administration's FY2017 budget requested $750 million for the fund ($500 million from the U.S. Department of State and $250 million from the U.S. Department of the Treasury).

U.S. Contributions

On March 8, 2016, the United States made an initial contribution of $500 million to the GCF. The funds were obligated with FY2016 budget authority from the State Department's Economic Support Fund—an account under the Function 150 and Other International Programs budget.

The Foreign Assistance Act of 1961, as amended (P.L. 87-195; part II, Chapter 4), provides the President authority to use the Economic Support Fund "to furnish assistance to countries and organizations, on such terms and conditions as he may determine, in order to promote economic or political stability."

On January 17, 2017, the State Department announced that it made an additional $500 million grant to support the GCF. The agency's press release stated that the funding was provided "from the fiscal year 2016 Economic Support Fund (ESF) appropriation."

The Trump Administration's FY2018 budget request, released on May 23, 2017, "eliminate[d] U.S. funding for the Green Climate Fund (GCF) in FY2018, in alignment with the President's promise to cease payments to the United Nations' climate change programs." Subsequent budget requests have also eliminated funding.

ISSUES FOR CONGRESS

Members of Congress hold mixed views about the value of international financial assistance to address climate change. While some Members are convinced that human-induced climate change is a high-priority risk that must be addressed through federal actions and international cooperation, others are not in agreement. Some are wary, as well, of international processes that could impose costs on the United States, redirect funds from domestic budget priorities, undermine national sovereignty, or lead to competitive advantages for other countries. Nonetheless, as a party to the UNFCCC, the United States has certain obligations under the treaty. The executive branch continues to negotiate and implement the UNFCCC obligations. Committees of Congress engage in oversight (from home and at the COP), provide input to the executive branch formally and informally, and decide program authorities and appropriations for these activities.

As Congress considers potential authorization and/or appropriations for the GCF (or, conversely, their prohibition), concerns may arise regarding the fund's cost, purpose, direction, efficiency, and effectiveness. Further, Congress might weigh the allocation of foreign assistance funds between bilateral and multilateral channels, as well as among the variety of multilateral mechanisms. Congress may also give guidance to the Departments of State and the Treasury with respect to the GCF's internal policies and external commitments. Further, Congress may gauge the impacts that U.S. participation in the fund might have on environmental protection, private sector investment, technological innovation, national security, humanitarian efforts, and international leadership.

For more discussion on the benefits and costs of international environmental assistance, see CRS Report R41845, The Global Climate Change Initiative (GCCI): Budget Authority and Request, FY2010-FY2016.

In: Climate Change
Editor: Jeremy Nesbitt
ISBN: 978-1-53617-216-4

Chapter 8

Paris Agreement: U.S. Climate Finance Commitments*

Richard K. Lattanzio

International Environmental Assistance

Many governments hold that environmental degradation and climate change pose international and trans-boundary risks to human populations, economies, and ecosystems. To confront these challenges, governments have negotiated various international agreements to protect the environment, reduce pollution, conserve natural resources, and promote sustainable growth. While some observers call upon industrialized countries to take the lead in addressing these issues, many recognize that efforts are unlikely to be sufficient without similar measures being taken in lower- income countries.

However, lower-income countries, which face poverty and economic growth issues, may not have the financial resources, technological know-how, and/or institutional capacity to deploy environmentally protective

* This is an edited, reformatted and augmented version of Congressional Research Service Publication No. IF10763, Updated October 2, 2019.

measures on their own. Therefore, international financial assistance, or foreign aid, has been a principal method for higher-income governments to support actions on global environmental problems in lower-income countries. Often, this assistance can serve as a cost-effective strategy for donor countries to provide greater market access for their environmental goods and services abroad and increased environmental benefits at home.

The United States and other industrialized countries have committed to providing financial assistance for global environmental initiatives through a variety of multilateral agreements, including, among others, the Montreal Protocol (1987), the United Nations Framework Convention on Climate Change (UNFCCC, 1992), and the U.N. Convention to Combat Desertification (1994). International financial assistance takes many forms, from fiscal transfers to market transactions. It may include grants, loans, loan guarantees, export credits, insurance products, and private sector investment. It may be structured as official bilateral development assistance or as contributions to multilateral development banks and other international financial institutions.

Ultimately, U.S. government assistance to developing countries for environmental initiatives is determined by Congress. Congressional committees of jurisdiction include the House Committees on Foreign Affairs, Financial Services, and Appropriations and the Senate Committees on Foreign Relations and Appropriations.

Finance Under the UNFCCC

The UNFCCC is the principal international treaty to acknowledge and address human-driven climate change. The United States ratified the treaty in 1992 (U.S. Treaty Number: 102-38).

Among the obligations outlined in Article 4 of the Convention, higher-income Parties (i.e., those listed in Annex II of the Convention, which were members of the Organization for Economic Cooperation and Development in 1992) sought to provide unspecified amounts of "new and additional financial resources to meet the agreed full costs incurred by developing country Parties in complying with their obligations" under the Convention. Further, "the implementation of these commitments shall take into account the need for adequacy and predictability in the flow of funds and the importance of appropriate burden sharing among the developed country Parties."

Over the past several decades, the United States has delivered financial and technical assistance for climate change activities in the developing world through a variety of bilateral and multilateral programs. U.S.-sponsored bilateral assistance has come through programs at the U.S. Agency for International Development, the U.S. State Department, the Millennium Challenge Corporation, the Export-Import Bank, and the Overseas Private Investment Corporation, among others.

U.S.-sponsored multilateral assistance has come through contributions by the U.S. Departments of State and the Treasury to environmental funds at various international financial institutions and organizations such as the Global Environment Facility (GEF), the Green Climate Fund (GCF), the U.N. Development Program, the U.N. Environment Program, the UNFCCC's Special Climate Change Fund and Least Developed Country Fund, and the World Bank's Climate Investment Funds and Forest Carbon Partnership Facility, among others. Each channel has its own mission and particular capacities.

Global Environment Facility

To facilitate the provision of climate finance under the UNFCCC, the Convention establishes a financial mechanism to provide funds to developing country Parties. Article 11 of the Convention states that the operation of the financial mechanism can be entrusted to one or more existing international entities.

The GEF was established in 1991 as an independent international financial institution to provide grants, promote cooperation, and foster actions in developing countries to protect the global environment. The GEF subsequently became an official operating entity of the financial mechanism of several international environmental agreements, including the UNFCCC. The relationship between the UNFCCC and the GEF is outlined in a memorandum of understanding contained in UNFCCC Conference of Parties (COP) decisions 12/CP.2 and 12/CP.3.

The George H. W. Bush Administration supported the establishment of the GEF in 1991. The United States has made commitments to all six GEF resource replenishments, including $430 million in 1994, $430 million in 1998, $430 million in 2002, $320 million in 2006, $575 million in 2010, $546 million in 2014, and $273 million in 2018, for a total of just over $3.0 billion. U.S. commitments correspond to about 12% of GEF's total. U.S. disbursements to the GEF since 1994 have totaled approximately $2.7 billion.

FINANCE UNDER THE CANCUN AGREEMENT

The 2009 COP in Copenhagen, Denmark, took note of a non-legal political document called the Copenhagen Accord. The following year, in Cancun, Mexico, the COP officially adopted many of the accord's elements into the Cancun Agreements (CA), including several quantified financial arrangements. The CA puts forth a collective commitment by developed country Parties (not specified in the text) to seek to provide new and additional resources approaching $30 billion for the period 2010-2012 to address the climate finance needs of developing countries (1/CP.16§95). The CA takes note of the pledge by developed country Parties to achieve a goal of mobilizing jointly $100 billion per year by 2020 to address the climate finance needs of developing countries (1/CP.16§98). The CA states that "funds provided to developing country Parties may come from a wide variety of sources, public and private, bilateral and multilateral, including alternative sources" (1/CP.16§99).

Finance under the Paris Agreement

In 2015, the COP in Paris, France, adopted the Paris Agreement (PA). Article 9 of the PA reiterates the obligation in the Convention for developed country Parties to seek to mobilize financial support to assist developing country Parties with climate change mitigation and adaptation efforts (Article 9.1). Also, for the first time under the UNFCCC, the PA encourages all Parties to provide financial support voluntarily, regardless of their economic standing (Article 9.2). The agreement states that developed country Parties (not specified in the text) should take the lead in mobilizing climate finance and that the mobilized resources may come from a wide variety of sources. It adds that the mobilization of climate finance "should represent a progression beyond previous efforts" (Article 9.3). The COP decision to carry out the PA (1/CP.21) uses exhortatory language to restate the CA's collective pledge by developed countries of $100 billion annually by 2020 and calls for continuing this collective mobilization through 2025. In addition, the Parties agree to set, prior to their 2025 meeting, a new collective, quantified goal for mobilizing financial resources of not less than $100 billion annually to assist developing country Parties.

The Green Climate Fund

The GCF is an official operating entity of the financial mechanism of the UNFCCC. The fund was proposed during the 2009 COP in Copenhagen, Denmark, and its design was agreed to by all Parties during the 2011 COP in Durban, South Africa. The GCF was made operational in 2014. The GCF aims to assist lower-income countries in their efforts to combat climate change through the provision of grants and other concessional financing for mitigation and adaptation projects, programs, and activities. The GCF is capitalized by "financial inputs from developed country Parties to the Convention" and "may also receive financial inputs from a variety of other sources, public and private, including alternative sources" (3/CP.17§§A29-A30).

The GCF was officially opened for capitalization at the U.N. Climate Summit in September 2014. Parties called for an immediate capitalization of between $10 billion and $15 billion over the course of the first year. Initial funding came from Germany, France, and a dozen other countries that pledged approximately $2.3 billion. Further pledges brought the total to approximately $10 billion.

U.S. Commitments

The Obama Administration announced a U.S. pledge of $3 billion over four years during the G-20 meetings in Australia on November 15, 2014.

U.S. Contributions

Under the Obama Administration, the U.S. State Department made two separate contributions of $500 million to the GCF on March 8, 2016, and on January 17, 2017. The funds were obligated with FY2016 budget authority from the agency's "Economic Support Fund" account. No contribution was made for FY2017.

THE TRUMP ADMINISTRATION

The Trump Administration's FY2018 budget request, released on May 23, 2017, proposes to "eliminate U.S. funding for the Green Climate Fund (GCF) in FY2018, in alignment with President Trump's promise to cease payments to the United Nations' climate change programs." Notwithstanding, the U.S. disbursement of $1 billion to the GCF under the Obama Administration ensures that the United States remains a member of the GCF board at least through the end of the initial phase of capitalization.

On June 1, 2017, President Trump announced his intention to withdraw from the PA. Under the provisions of the PA, this could not be completed before November 4, 2020.

While a withdrawal from the PA would release the United States from specific obligations under the PA, it would not release the United States from the obligations under the UNFCCC to provide "new and additional financial resources to meet the agreed full costs incurred by developing country Parties in complying with their obligations" under the Convention or the collective pledge by developed country Parties to achieve "a goal of mobilizing jointly $100 billion per year by 2020."

The effects of the Trump Administration's announced withdrawal from international climate finance commitments would rely on assumptions regarding the efficiencies of these programs and the sum of the collective assistance provided by other nations, regional governments, cities, and the private sector. Some analysis suggests the impacts could be minimal if other contributions compensate for the lack of U.S. assistance. However, the U.S. decision could also lead to an erosion of international efforts to help developing countries confront climate change.

In: Climate Change
Editor: Jeremy Nesbitt

ISBN: 978-1-53617-216-4

Chapter 8

POTENTIAL IMPLICATIONS OF U.S. WITHDRAWAL FROM THE PARIS AGREEMENT ON CLIMATE CHANGE*

Jane A. Leggett

On June 1, 2017, President Trump announced his intent to withdraw the United States from the Paris Agreement (PA), an international accord to address climate change over the coming century. In a December 2018 meeting of the Parties, an official U.S. statement reiterated the intent to withdraw "absent the identification of terms that are more favorable to the American people."

Congress may wish to consider implications of the President's intent in several dimensions: foreign policy, economic, environmental, and legal.

---

* This is an edited, reformatted and augmented version of Congressional Research Service Publication No. IF10668, updated April 5, 2019.

WHAT IS THE PARIS AGREEMENT?

The PA exists under the United Nations Framework Convention on Climate Change (UNFCCC). The United States ratified the UNFCCC in 1992 with the advice and consent of the Senate. In 2016, President Obama accepted the PA without requesting the Senate's advice and consent; the Department of State took the view that the PA contained no substantive, legal obligations for the United States beyond those already required by the UNFCCC. Currently, the PA has 185 Parties—governments that have ratified or accepted the agreement, including the United States—out of 195 Signatories. Of the top 20 emitting nations, only Iran, Russia, and Turkey are not Parties.

Under the UNFCCC, the United States and the then-industrialized Parties listed in "Annex I" took on specific commitments for greenhouse gas (GHG) mitigation, financial resources, reporting, and review. The PA modified the UNFCCC bifurcation of commitments between Annex I and developing country Parties to establish a single, common set of obligations for all Parties (with flexibilities for the least capable). For example, China agreed in the PA to the same binding commitments as the United States.

Under the PA, all Parties must submit "Nationally Determined Contributions" (NDC) every five years, but the content is not binding. An NDC must identify how the Party intends to abate its GHG emissions, initially to 2025 or 2030, depending on the time frame each nation chooses. Each nation voluntarily decides its NDC. As such, all PA emissions targets, including the U.S. target, are voluntary and nonbinding.

The UNFCCC requires GHG and other reporting and review, differentiated by types of Parties. The PA makes an enhanced transparency framework applicable to all Parties. The PA contains cooperative compliance mechanisms but not formal sanctions. Incentives for compliance include "name and shame" processes: Parties that do not meet their pledges may incur diplomatic and public opinion penalties. For example, the European Union has a policy not to sign trade pacts with countries that are not parties to the PA.

Variety of GHG Pledges in Selected NDCs United States

Reduce GHG emissions to 26-28% below 2005 levels in 2025.

China: Inter alia, by 2030, peak carbon dioxide (CO_2) emissions; lower its CO_2 emissions per unit of gross domestic product (GDP) by 60-65% below 2005 levels; and increase the non-fossil- fuel share of energy consumption to around 20%.

European Union: Reduce GHG emissions by at least 40% compared to 1990 levels.

India: Inter alia, reduce GHG emissions per unit of GDP by 33- 35% by 2030 from the 2005 level and reach 40% of its cumulative installed electric capacity from non-fossil-fuel sources by 2030.

Mexico: Reduce GHG and black carbon emissions by 25% below business-as-usual projections by 2030, and up to 40% under some conditions, implying a net emissions peak in 2026. Reduce GHG emissions per unit of GDP by around 40% from 2013 to 2030.

Countries have long negotiated over what would constitute a "fair" distribution of effort under the UNFCCC. The strength of the U.S. NDC compared with those of other countries can be viewed from various perspectives (see text box): To illustrate, under the NDCs, China's GHG emissions could grow to 2030, while U.S. GHG emissions would fall. Nonetheless, China's GHG emissions per person would remain well below those of the United States, and China would reduce its emissions per unit of GDP more than the United States.

WITHDRAWAL PROCEDURE

The Department of State notified the U.N. Secretary General that the United States would provide formal notification of withdrawal "as soon as it is eligible to do so." Under Article 28 of the PA, this would be November 4, 2019. Withdrawal may take effect one year later—on or after November 4, 2020. In the meantime, the United States remains a Party to the PA (unless, following customary international law, the other Parties agree to allow an earlier exit).

In the meantime, the U.S. delegation continues to participate "in order to ensure a level playing field that benefits and protects U.S. interests," according to the Department of State.

FOREIGN POLICY AND DIPLOMACY

The President's announcement was viewed generally by observers as consistent with the Administration's "America First" approach to foreign policy. Although the Administration has arguably sent mixed signals about its specific foreign policy intentions, some of these signals appear to reflect, among other things, a skeptical approach toward multilateral organizations and multilateral agreements and a transactional approach to alliances and international agreements.

Some observers argue that the Administration's decision to withdraw from the PA will (1) reduce the U.S. standing in the world by making the United States an international outlier on climate change, (2) strengthen perceptions that the United States is withdrawing from its traditional position of world leadership and becoming more inward- focused or even isolationist, (3) create an opportunity for China to assume a position of world leadership on climate change and perhaps other issues, and (4) make the United States appear less reliable as a negotiating partner, which could make it harder for the United States in the future to secure foreign cooperation for addressing other issues of mutual interest or to call on other countries to abide by their commitments in other international agreements. Other observers either disagree with these arguments or argue that they are offset by gains realized for Americans under the Administration's approach to foreign policy.

Other nations responded to the U.S. intent to withdraw largely with restrained expressions of regret; they rebuffed the President's proposal to reopen negotiations. Despite initial concerns about the resolve of Parties to implement the PA without the United States, the most recent meeting of the Parties agreed on most of the "rulebook" for implementing the PA's provisions. China's Special Representative on Climate Change stated, just prior to those negotiations, that "the political influence of the American withdrawal [from the PA] was quite big." He said that the U.S. announcement initially "affected the resolve and confidence of some other countries." He opined that the impact of the U.S. decision has since "dissipated," in part because of China's pledge to meet its commitments in full. China "sent out such a strong political signal," he said, that it helped "stabilize" international climate change efforts.

ECONOMY AND TRADE

The PA was intended to lay a path to long-term transition of the world's economies toward "deep decarbonization"— sustaining economic growth while delinking it from emissions of CO2 and other GHGs. Such a transition is generally expected to impose near- to medium-term costs, though the magnitude and distribution would depend on policy timing and design and private investment. Trade could be affected by differences in Parties' NDCs and policy designs. Economic and trade impacts could be adverse or positive for individual countries or businesses, and many are building this into their economic strategies.

Experts expect that deep decarbonization would initially increase costs and dislocation for many populations and businesses. Technology experts and economists expect that experience with and wider use of advanced technologies would lower the costs of key GHG mitigation options, and economies would adjust to new conditions.

Some economists believe that the costs of mitigating GHG not fully mitigate adverse impacts. Optimizing policy might seek to minimize the total net costs of mitigation, adaptation, and losses and consider distributional impacts.

The trade implications of the President's intent to withdraw are unclear. The President's Executive Order 13783, Promoting Energy Independence and Economic Growth, instructed agencies to review, and possibly rescind or modify, regulations that may obstruct energy development, including those that would reduce U.S. GHG emissions. As China, South Korea, and European and other countries enact climate change policies, they promote investment in advanced energy, materials, electronics, vehicle, and other technologies. They may initially raise costs, but over time, they expect many technologies to become commercially competitive and bring trade advantages. To date, the U.S. Congress has largely continued public funding to advance cleaner technologies, but the level of private sector research and "learning-by-doing" is uncertain with lower policy stimulus for technology end-users.

ENVIRONMENT

The PA contains a collective commitment to reduce GHG emissions almost to net zero in the second half of this century. The goal is to hold the GHG-induced increase in global temperature well below 2o Celsius and to try to hold it to 1.5oC. Reaching "net zero" would stabilize the rise of GHG concentrations in the atmosphere and, after a lag of many decades, end the induced climate change. The pledges made in NDCs to 2025 or 2030 are intended as the initial steps in a multi-decadal process toward the net zero goal.

President Trump cited one study estimating that the initial NDCs would avoid global warming of about 0.2oC by 2100. This could be roughly 8-10% of the way to meeting the 2oC goal. Compared with a no-further-policy scenario, some of the study's authors estimated that countries' commitments could reduce temperature by 0.6-1.1° C. Other studies estimate larger or smaller temperature effects of NDCs, depending on assumptions about future commitments.

LONG-TERM LIABILITIES

President Trump cited "legal liability" associated with staying in the PA. This may refer to an assertion that PA obligations could prevent the federal government from rescinding or modifying its domestic regulations or risk litigation under the Clean Air Act. In rebuttal, a former Department of State attorney who worked on the PA has opined that "if a domestic stakeholder sought to invoke the PA in a domestic challenge to withdrawing the Clean Power Plan, courts would almost certainly find that the agreement does not constrain executive branch action." The EPA has since proposed to rescind the Clean Power Plan. Thus far, references to commitments in the PA have not featured in litigation challenging the Clean Power Plan or comments submitted on EPA's proposal to revise vehicle GHG emission standards.

In: Climate Change
Editor: Jeremy Nesbitt
ISBN: 978-1-53617-216-4

Chapter 9

CLIMATE CHANGE AND THE AMERICA'S TRANSPORTATION INFRASTRUCTURE ACT OF 2019 (S. 2302)*

William J. Mallett

Highway, public transportation, and rail programs are authorized through FY2020 by the Fixing America's Surface Transportation (FAST) Act (P.L. 114-94). In August 2019, the Senate Committee on Environment and Public Works reported a bill that would reauthorize the highway elements of surface transportation programs from FY2021 through FY2025. The bill, S. 2302, the America's Transportation Infrastructure Act of 2019, would be the first surface transportation authorization act to include major provisions that address climate change.

Surface transportation is a major source of carbon dioxide (CO2) in the atmosphere, the main human-related greenhouse gas (GHG) contributing to climate change. At the same time, the effects of climate change, such as extreme heat, sea level rise, and stronger storms, pose a threat to transportation infrastructure. S. 2302 seeks to address these two aspects of

* This is an edited, reformatted and augmented version of Congressional Research Service Publication No. IF11299, dated September 3, 2019.

climate change with mitigation provisions that aim to reduce GHG emissions from surface transportation and adaptation provisions that aim to make the surface transportation system more resilient to a changing climate.

Climate Change Mitigation

Transportation is the largest source of GHG emissions in the United States, and thus is a target of emission reduction efforts to mitigate future climate change. The Environmental Protection Agency (EPA) estimates that, in 2017, transportation accounted for 29% of U.S. man-made GHG emissions. GHG emissions from the transportation sector come mainly from cars and light trucks (Figure 1). Total GHG emissions from transportation increased by 22% between 1990 and 2017, but peaked in 2005. Total emissions were about 6% less in 2017 than in 2005, but have increased since 2012. Most GHG emissions from the sector are due to the release of CO2 from the combustion of gasoline and diesel fuel. Other GHGs, such as methane, make up about 3% of emissions from transportation.

GHG emissions from transportation are largely a function of three factors: vehicle fuel efficiency, the carbon content of fuel, and vehicle miles traveled. Vehicle fuel efficiency is regulated by the Corporate Average Fuel Economy (CAFE) standards, administered by the Department of Transportation's National Highway Traffic Safety Administration, and GHG emissions standards, administered by EPA. The CAFE standards are established under the authority of the Energy Policy and Conservation Act, as amended, and the GHG standards under the authority of the Clean Air Act, as amended. Neither is commonly legislated in surface transportation authorizations.

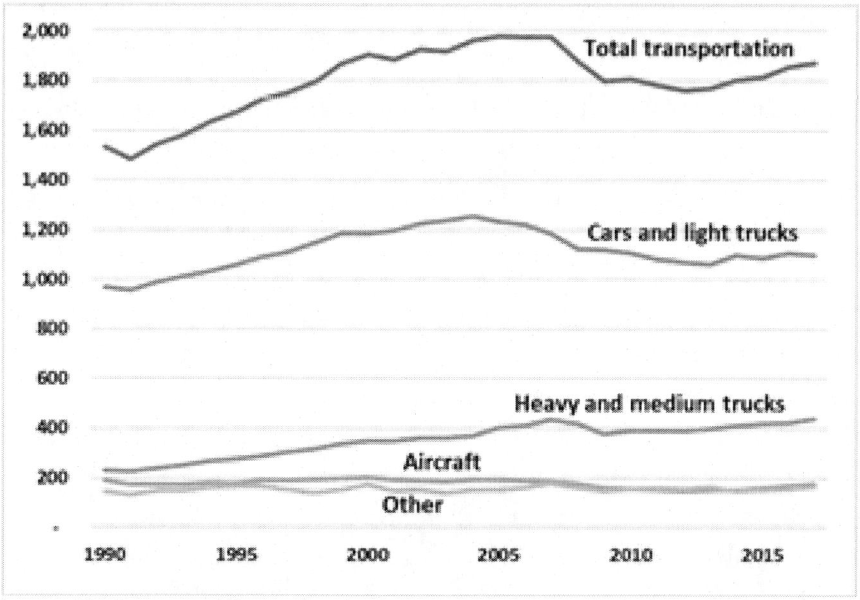

Source: Environmental Protection Agency.
Note: "Other" includes buses, motorcycles, ships and boats, rail, and pipelines.

Figure 1. Greenhouse Gas Emissions from Transportation in the United States, 1990-2017 Millions of metric tons of carbon dioxide equivalent.

Other climate mitigation policies affecting surface transportation are generally the indirect effects of policies enacted for other reasons. For example, the Congestion Mitigation and Air Quality Improvement (CMAQ) program, part of the Federal-Aid Highway Program, provides federal funding for projects that contribute to the attainment of ambient air pollution standards for ozone, carbon monoxide, and particulate matter. This typically involves projects that reduce pollutant emissions from cars and trucks that also co-emit CO_2. Other surface transportation programs that may contribute indirectly to the reduction of GHG emissions include the Transportation Alternatives program, which funds projects such as bicycle and pedestrian infrastructure, and the federal public transportation program.

Climate Change Adaptation

Climate change is likely to include higher average temperatures, greater extremes of temperature, more precipitation overall with an increase in precipitation intensity and greater variation, and a rise in sea level. Climate change may also lead to fewer but stronger hurricanes. While the consequences of some of these changes may depend to some extent on other human activities, such as urban development patterns, they are likely to include an increase in extreme heat; fewer days below freezing; more coastal, riverine, and flash flooding; and more droughts and wildfires. Intense precipitation could lead to more mudslides, particularly following droughts and wildfires.

Existing surface transportation infrastructure can be vulnerable to climate change because it was constructed for sea level and weather extremes that are being or are likely to be exceeded in the future. If the effects of climate change worsen, as studies anticipate, the impacts of extreme weather on surface transportation infrastructure and operations are likely to increase in magnitude, duration, and frequency. For example, an increase in the number of very hot days may cause more damage to bridges because of greater thermal expansion of bridge joints. More intense precipitation and flooding could result in more road washouts, bridge scour, and roadside mudslides. Not all the effects of climate change will be negative. For example, a warmer climate could reduce road pavement deterioration in some places due to less freezing, snow, and ice.

Adaptation is action to reduce the vulnerabilities and increase the resilience of the transportation system to the effects of climate change. Adaptation options for surface transportation include structural and nature-based engineering and policy-based activities. For example, highway bridges can be engineered structurally to withstand the threats of higher wind and water. Nature-based engineering may involve reducing climate vulnerabilities through activities such as wetland restoration, artificial reefs, and beach restoration. Policy-based activities include changing maintenance practices, such as more frequent cleaning of drains, and improving operations plans for weather emergencies.

Currently, there is no dedicated surface transportation funding for adaptation projects. The Federal Highway Administration (FHWA) has stated that federal-aid highway funds can be used to assess the potential impacts of climate change and to apply adaptation strategies. FHWA's Emergency Relief program, which provides funds for rebuilding after natural disasters, also allows some spending on resiliency features. Moreover, several aspects of federal law, regulation, and policy require asset managers to consider the effects of climate change on surface transportation infrastructure. FHWA, in cooperation with state departments of transportation, has sponsored vulnerability assessments and conducted research into making surface transportation more resilient to climate change.

America's Transportation Infrastructure Act of 2019 (S. 2302)

S. 2302 would authorize about $10 billion over five years specifically for highway-related climate change mitigation and adaptation programs. More funding could be available for mitigation and adaptation activities if state and local governments choose to use a portion of other federal highway grants for this purpose, as S. 2302 would allow. Funding would be authorized from the Highway Trust Fund, which derives most of its revenue from a tax on motor vehicle fuels. The major climate change-related provisions proposed by S. 2302 include the following:

Mitigation Provisions

Establishes new formula and competitive grant programs to support planning and projects that reduce on-road mobile sources of CO2 emissions. Eligible projects might include ridesharing programs, truck stop electrification, and incident management programs.

Funding would average $600 million annually for the formula program and $100 million annually for the competitive program.

- Establishes a new competitive grant program to support construction and operation of alternative fueling infrastructure along designated alternative fuel corridors. Funding would average $200 million annually.
- Establishes a new competitive grant program to support projects that would reduce GHG and air pollutant emissions at ports by reducing truck idling. Funding would average $74 million annually.
- Establishes a federal interagency working group to develop a strategy to transition the vehicle fleets of federal agencies to hybrid-electric vehicles, plug-in electric drive vehicles, and alternative fueled vehicles.
- Establishes a new competitive grant program for highway congestion reduction projects that may indirectly reduce transportation emissions. Funding would average $40 million annually.

ADAPTATION PROVISIONS

- Establishes a new grant program to support adaptation projects. Funding would average $986 million annually, with $786 million distributed to the states by formula and $200 million distributed competitively. The program would also encourage the development of resilience improvement plans.
- Makes certain "protective features" designed to mitigate the risk of recurring damage from extreme weather events, flooding, or other natural disasters eligible expenses under the federal highway program. The federal government would pay up to 100% of the cost of projects such as raising roadway grades, stabilizing slopes, and adding bridge scour protection; for most other types of

highway construction, the states must pay at least 10% or 20% of the cost.

- Adds adaptation strategies to the required contents of the National Freight Strategic Plan and state freight plans.
- Adds wildfire and sea level rise to the definition of natural disaster in the Emergency Relief program, and explicitly makes economically justifiable resilience features eligible for funding as part of repair and reconstruction projects.
- Defines "natural infrastructure" as infrastructure that uses, restores, or emulates natural ecological processes, and makes such projects specifically eligible for funding.

In: Climate Change
Editor: Jeremy Nesbitt
ISBN: 978-1-53617-216-4

Chapter 10

A BRIEF COMPARISON OF TWO CLIMATE CHANGE MITIGATION APPROACHES: CAP-AND-TRADE AND CARBON TAX (OR FEE)*

Jonathan L. Ramseur

Almost all climate scientists agree that greenhouse gas (GHG) emission increases have contributed to observed climate change and that continued increases in GHG emissions will contribute to future climate change.

Although a variety of efforts seeking to reduce GHG emissions are currently underway on the international level and in individual states or regional partnerships, federal policymakers and stakeholders have different viewpoints over what to do—if anything—about GHG emissions, future climate change, and related impacts.

For policymakers considering actions to reduce GHG emissions, various policy instruments are available. Over the last 15 years, many

* This is an edited, reformatted and augmented version of Congressional Research Service Publication No. IF11103, dated February 12, 2019.

legislative proposals have involved market-based approaches, such as a GHG emissions cap- and-trade system or a carbon tax. These particular approaches may be considered in the 116th Congress and are discussed below. The information below provides an overview of two approaches while briefly addressing their similarities and differences.

What Is a Cap-and-Trade System?

A cap-and-trade system is a policy tool that creates a cap on GHG emissions from selected emission sources while providing the sources with flexibility—on-site reduction or emissions trading—when complying with the emissions cap. The cap could apply to the primary GHG emitted by human activity, carbon dioxide (CO2), or it could apply to multiple GHGs, such as methane, nitrous oxide, or fluorinated gases. Covered sources in prior legislative proposals have included major emitting sectors (such as power plants and specific industries), fuel producers and/or processors (such as coal mines or petroleum refineries), or some combination of both.

An emissions cap is partitioned into emission allowances (or permits). Typically, in a GHG cap-and-trade system, an emission allowance represents the authority to emit one metric ton of CO2-equivalent—a measure that accounts for different GHG global warming potentials.

Policymakers may decide to (1) sell the emission allowances through periodic auctions, which would generate a new federal revenue stream; (2) distribute allowances to covered sources at no cost (based on, for example, previous years' emissions); or (3) use some combination of these strategies. Given that emission allowances have a market value, the distribution of emission allowances would likely be a source of significant debate during a cap-and-trade program's development, as discussed below.

At the end of each established compliance period (a calendar year or multiple years), covered sources submit emission allowances to an implementing agency to cover the number of tons emitted during the period. Generally, if a source did not provide enough allowances to cover its emissions, the source would be subject to penalties.

Under an emissions cap, covered sources would have a financial incentive to make reductions beyond what is required, because they could (1) sell unused emission allowances to entities that face higher costs to reduce their facility emissions, (2) reduce the number of emission allowances they need to purchase, or (3) bank emission allowances—if allowed—to use in a future compliance period.

A cap-and-trade system would create an emissions trading market. Depending on program design details, emission allowance trading could involve not only sources directly subject to an emissions cap but also a range of brokers and intermediaries. The federal government oversees existing emissions trading programs (sulfur dioxide and nitrogen oxides) and would likely oversee a GHG program.

WHAT IS A CARBON TAX (EMISSIONS FEE)?

A carbon tax or emissions fee is a policy tool that provides a financial incentive to reduce GHG emissions by attaching a price to GHG emissions (CO2 emissions or multiple GHGs) or their emission inputs, namely fossil fuels. The choice of terminology between a tax or fee may have procedural consequences, particularly in terms of congressional committee jurisdiction, which could potentially influence the policy's design. As many policymakers, stakeholders, and academic journals use the term carbon tax, this is the default term in this document.

A central policy choice when establishing a price on GHG emissions is the rate of the carbon tax (measured in dollars per ton of emissions). Several factors could be considered when setting the rate. For example, Congress could set the rate at a level or pathway—based on modeling estimates— that would achieve a specific GHG emissions target.

Congress may also consider whether the tax rate should increase over time and, if so, by how much.

A carbon tax would generate a new revenue stream. The magnitude of the revenues would depend on the scope and rate of the tax, the responsiveness of covered entities in reducing their potential emissions, and multiple other market factors. A 2018 Congressional Budget Office study estimated that a \$25/metric ton tax on energy-related and other GHG emissions would yield approximately \$100 billion each year during the first 10 years of the program.

The distribution of this new revenue stream would likely be a source of significant debate, as discussed below.

Similarities Between Approaches

Cap-and-trade and carbon tax instruments are market-based approaches that may be used to reduce GHG emissions. In many ways, a cap-and-trade program and carbon tax would produce similar effects. Both would place a market price on GHG emissions (directly or indirectly), and both would increase the relative market price of more carbon-intensive energy sources, particularly coal, which generate greater emissions per unit of energy. This result could lead to the displacement of these sources with lower carbon-intensive sources, including renewables; spur innovation in emission reduction technologies; and stimulate actions that may decrease emissions, such as efficiency improvements.

Distribution of Allowance Value and Tax Revenue

When designing either program, policymakers would likely face challenging decisions regarding the distribution of the new emission allowance value (which includes both auction revenues and distribution of no-cost allowances) or tax revenues. Policymakers could apply the allowance value or tax revenue to support a range of policy objectives but would encounter trade-offs among objectives.

The central trade-offs include minimizing economy-wide costs (often measured in terms of gross domestic product); lessening the costs borne by specific groups, particularly low-income households and displaced workers or communities; and supporting a range of specific policy objectives, which may or may not be related to climate change.

ECONOMIC IMPACTS

A primary concern with either approach regards their potential for economy-wide impacts and disproportionate effects on particular demographic groups and specific industries. The degree of these potential effects would depend on multiple factors, including the scope of the program and the use of allowance value or tax revenues.

Policymakers may have different perspectives on whether estimated economy-wide costs—often measured in terms of U.S. gross domestic product—are significant. In addition, some would argue that these costs be compared with the climate benefits achieved from the program as well as the estimated costs of taking no action. Estimates of climate- related benefits and costs often contain considerable uncertainty and have generated debate in recent years.

Either approach may yield disproportionate impacts on certain demographic groups, particularly lower-income households, which spend a greater proportion of their income on energy needs. Congress could address these impacts by distributing some of the allowance value or tax revenues back to households in some fashion.

While some domestic industries may thrive, a price on GHG emissions could create a competitive disadvantage for other industries, particularly emission-intensive, trade- exposed industries. Policymakers could address some of the potential concerns, for example, by including a border adjustment mechanism and allocation of no-cost emission allowances or tax rebates to selected industries.

Coverage Decisions

Under either approach, policymakers would face a similar debate regarding scope and applicability. For example, questions such as which emission sources should be subject to the program or which GHG emissions to include would be raised with either approach. Policymakers may consider multiple factors when debating these issues, including environmental effectiveness, economic efficiency, costs, measurement, available technology, and administrative concerns.

Role of Emission Offsets

Many existing cap-and-trade programs allow for the use of emission offsets as a compliance option. In a carbon tax program, policymakers could allow for tax credits for offset-type projects. An offset is a measurable reduction, avoidance, or sequestration of GHG emissions from a source not covered by an emission reduction program.

Economic analyses have found that offset treatment could have a substantial impact on overall program cost, because these projects can often reduce emissions at a lower cost than many typically covered sources. However, in existing cap-and-trade programs, offsets have generated some controversy and raised concerns, including the credibility of emission reductions from particular offset projects and environmental justice issues more generally.

Price Control Versus Quantity Control

A primary difference between a cap-and-trade system and a carbon tax program is that the former provides emissions certainty, while the latter provides price certainty. In one sense, preference for a price (carbon tax) or a quantity limit (emissions cap) depends on one's preference for uncertainty—either uncertain emissions or uncertain program costs.

Policymakers can include multiple design elements, such as a price safety valve or auction reserve price, with a cap-and-trade program that may blur the distinction between price and quantity control. Similarly, a carbon tax program could include a mechanism by which policymakers could alter the tax rate if they determine that emission reductions are not proceeding at a desirable pace.

CONCLUDING OBSERVATIONS

Discussions of cap-and-trade and carbon tax approaches often center on their potential advantages in terms of emissions uncertainty and price uncertainty, respectively. The degree to which one approach has an advantage in a particular context, such as transparency or administration, would depend on the designs of the programs being assessed. In many cases, these differences may be addressed with specific design elements.

Although recent attention has largely focused on market- based mechanisms, policymakers can address emissions with other policy tools, including performance-based regulations, which currently apply to motor vehicle emissions, or promotion of mitigation technologies, such as carbon capture and sequestration. These tools could support market-based programs or function independently.

In: Climate Change
Editor: Jeremy Nesbitt

ISBN: 978-1-53617-216-4

Chapter 11

PROJECTED ECONOMIC IMPACTS OF CLIMATE CHANGE*

Danielle A. Arostegui and Jane A. Leggett

As mandated by Congress, the U.S. Global Change Research Program produces regular National Climate Assessment (NCA) reports on the state of scientific knowledge about climate change and its effects on human and natural systems in the United States. Volume II of NCA4, published in 2018, examines the contributions of individual economic studies, in order to develop consensus conclusions from across the body of climate impacts literature. According to NCA4, "annual losses in some economic sectors are projected to reach hundreds of billions of dollars by the end of the century" provided continued growth in emissions at historic rates. The findings of the NCA4, and those of previous reports, have raised concerns in Congress about the projected economic impacts and questions about the study methods used to derive these projections.

Studies projecting the economic impacts of climate change vary in their scope, methodology, and the intended applications of their findings.

---

* This is an edited, reformatted and augmented version of Congressional Research Service Publication No. IF11156, dated March 28, 2019.

Some focus on the effects of one aspect of climate change upon a single type of economic activity (e.g., effects of temperature and carbon dioxide fertilization on corn yields), while others study the effects across a more comprehensive array of market and non-market activities. Recent research, however, provides detailed projections of future impacts of climate change across a variety of sectors of the U.S. economy.

Projections of the economic impacts of climate change have been cited by some Members of Congress to advance climate-related legislation, and have raised questions and concerns about their interpretation and use in policymaking. This product provides an abbreviated overview of climate impacts research, recent projections from the literature, and considerations in interpreting these. In this context, climate impacts are defined as the projected physical and economic effects of climate change, both positive and negative. This product does not discuss the social cost of carbon or similar metrics, which estimate the net present costs associated with emitting additional greenhouse gases (see CRS In Focus IF10625, Social Costs of Carbon/Greenhouse Gases: Issues for Congress).

STATUS OF U.S. CLIMATE IMPACTS RESEARCH

While scientific understanding of the physical effects of climate change is supported by a large body of research, the methods for projecting the likely physical and economic effects for specific regions are relatively new and imprecise. Development of these methods remains an area of active research.

Methods of projecting the economic effects of climate change have been in development since the early 1980s. Early attempts to quantify more comprehensive economic effects largely did so on a global scale. Since the early 2000s, advancements in the underlying physical and economic research, data availability, and computing power have allowed the development of more granular approaches, capable of estimating the sum of a wider array of sectoral effects at the national, regional, or even county level. These methods are complex, drawing together insights from a range of disciplines, including climate science, economics, and statistics.

According to a 2017 report of the U.S. Government Accountability Office (GAO), a "small but growing number of researchers" are developing the methods used to project the potential economic effects of climate change within the United States. To date, these methods have been primarily applied to certain economic sectors for which data are readily available and evidence for the relationship between climate and impacts is considered robust. These include human health, labor, energy, agriculture, water resources, infrastructure, and coastal property. Currently, no studies purport to produce a comprehensive estimate of the effects of climate change on the entirety of the U.S. economy.

RECENT PROJECTIONS IN THE LITERATURE

Two studies, cited in Volume II of the NCA4, provide the most detailed projections to date of the economic effects of climate change across multiple sectors in the United States. These are:

- A 2017 report of the Climate Impacts and Risk Analysis (CIRA) project, coordinated by the U.S. Environmental Protection Agency (EPA) with input from other federal agencies; and
- A 2017 study by Hsiang et al., published in Science, and building upon the American Climate Prospectus, a 2014 report of the Rhodium Group.

These studies vary in methodology, economic sectors that are included, and choice of metric to report results.

Consequently, their findings are not directly comparable.

The 2017 CIRA report provides national and regional economic impact projections for the analyzed sectors in 2050 and 2090 under two emissions trajectories. The report analyzes 22 climate impact sectors within six broad categories: health, infrastructure, electricity, water resources, agriculture, and ecosystems. Results are presented in real 2015 dollars for each sector, but are not aggregated into a single economy-wide estimate. Under a high emissions scenario with limited adaptation, net losses in labor, extreme temperature mortality, and coastal property are estimated in the hundreds of billions of dollars per year by the end of the century. Of the 22 sectors analyzed, one is estimated to result in net benefits in 2090.

The Hsiang et al. study projects net climate impacts in six sectors—agriculture, crime, coastal property, energy, human mortality, and labor—under three emissions trajectories for 2080 through 2099. They report their findings as a percentage of gross domestic product (GDP). Annual net economic costs across the sectors are estimated to cost "roughly 1.2% of [GDP] per +1 °C on average," with estimates ranging from +0.1% (net benefit) to -1.7% (net cost) GDP at low levels of warming (1.5 °C) through -6.4% to -15.7% of GDP under a higher warming scenario (8 °C).

Selected Considerations in Interpreting Climate Impact Projections

Climate impact researchers and other experts caution that estimates should not be interpreted as predictions of actual future impacts. Instead, researchers suggest the estimates may provide insights into the potential direction and magnitude of impacts, depending on varying assumptions. Some critics suggest that the inherent uncertainties associated with projecting to the end of the century make the net costs of climate change in that timeframe fundamentally unknowable. Beyond these central concerns, additional considerations, discussed below, may be of interest to policymakers.

Analytical Challenges

The NCA4 notes that the research literature on physical and economic impacts in the United States remains incomplete in its coverage of the range and magnitudes of potential impacts. Challenges to quantification include, among others:

- The wide variety of economic sectors likely affected by climate change, the complexity of the effects, and the existence of complex feedbacks among sectors;
- Lack of metrics for monetizing some of the non-market effects (e.g., loss of biodiversity, ecosystem damages);
- The long timescales over which some greenhouse gases persist in the atmosphere;
- The difficulty of long-term projecting;
- Uncertainty surrounding possible climatic or societal tipping points, beyond which impacts may accelerate or become irreversible; and
- Uncertainty surrounding the magnitude and efficacy of future adaptation, which may reduce the economic effects of physical impacts (and incur its own costs).

Some observers suggest that these challenges have resulted in estimates that systematically underestimate the actual future costs associated with climate change. Both the Hsiang et al. and the CIRA study acknowledge their incomplete coverage and plan to expand it as the underlying research develops. By contrast, some argue that insufficient accounting for adaptation may overestimate the economic impacts on humans and ecosystems.

TREATMENT OF SOCIAL VALUES

Economic models are sensitive to assumptions that often contain judgments about the relative value society places on various outcomes. Examples of social values that may be included in climate models are time preference/discounting (i.e., the extent to which society values costs occurring in the present more than those occurring in the future), risk tolerance, and consideration of outcomes that affect social inequality. Hsiang et al. project that the valuation of the economic damages increases by a factor of between 1.3 and 4.6 if social aversion to inequality is included in the analysis. There is no consensus about the appropriate treatment of social preferences in economic impact estimates. Whether and how these values are accounted for influences both the final estimate and its interpretation for policymakers.

DISTRIBUTION OF ECONOMIC IMPACTS

Projections of aggregate national effects may obscure variations by region and socioeconomic distribution. Climate effects vary by geographical region depending on factors such as the current climate and climate-sensitivity of the locality. For instance, CIRA projects relatively moderate climate impacts in some sectors in the Northwest as compared to other U.S. regions. These variations may transfer value over time from some regions of the country to others. Hsiang et al. project that climate change is likely to generate a transfer of value from the southern and eastern portions of the United States to the northern and western regions. The same study estimates that the impacts could disproportionately affect low-income communities, tending to increase existing inequalities.

CHOICE OF REPORTING METRIC

The choice of metrics used to report potential economic impacts can influence the interpretation of the results. For example, GDP, as calculated by the U.S. government, measures the value of goods and services produced in the United States, but does not necessarily distinguish between costs and benefits. As a result, some effects of climate change, which could be considered adverse impacts (e.g., property loss to hurricanes), may yield an increase in GDP because they increase economic activity for relief and recovery, while the loss of assets may not be fully counted.

Some researchers (including Hsiang et al.) calculate net costs under their own methodology rather than that traditionally used to calculate GDP, but still report their findings as a percentage of GDP in order to place the estimated costs in the broader context of total economic productivity. The interpretation of these estimates is distinct from those that directly calculate the impacts on future GDP levels or growth due to the differential treatment of costs and benefits when calculating the final estimates.

In: Climate Change
Editor: Jeremy Nesbitt
ISBN: 978-1-53617-216-4

Chapter 12

CLIMATE-RELATED RISK DISCLOSURE UNDER U.S. SECURITIES LAWS*

Eva Su and Nicole Vanatko

In light of public concern over climate change, some stakeholders have asked to what extent publicly traded companies should disclose their climate-related risks. The Securities and Exchange Commission (SEC) requires publicly traded companies to disclose financial statements and certain other relevant business information in public filings, including annual and quarterly reports. While current SEC requirements do not address climate-related risks expressly, publicly traded companies must disclose such risks if they are "material" under federal securities laws.

* This is an edited, reformatted and augmented version of Congressional Research Service Publication No. IF11307, dated September 10, 2019.

FINANCIAL RISKS POSED BY CLIMATE CHANGE

Numerous organizations, shareholder groups, businesses, and financial regulators have recognized financial risks that climate change may pose to companies. Such climate- related risks commonly fall into two general categories:

- Physical risks: These risks include direct and indirect risks arising from extreme weather events and from longer-term shifts in climate patterns, including, for example, changes in water availability and food security. Physical risks have important implications for many companies' physical facilities, operations, transportation costs, supply chains, and employees.
- Transition risks: These risks arise from policy, legal, technology, and market changes as the world transitions to a lower-carbon economy, with potential financial or reputational effects on businesses. For example, a company may engage in efforts to reduce greenhouse gas emissions or otherwise respond to changing consumer behavior.

Although financial regulators have not traditionally focused on climate change, some financial regulators have begun to consider the economic impact of climate-related changes. For example, the Bank of England is incorporating climate change scenarios into certain prudential regulation stress testing processes. Similarly, the Bank of Canada recently added climate change to its list of the top economic risks facing the country's financial system, as noted in its 2019 Financial System Review. In the United States, Federal Reserve Chairman Jerome Powell has stated that climate events "have the potential to inflict serious damage on the lives of individuals and families, devastate local economies, and even temporarily affect national economic output and employment."

WHAT CLIMATE-RELATED RISK DISCLOSURES DO THE SEC CURRENTLY REQUIRE?

Federal securities law does not explicitly require publicly traded companies to disclose specific climate-related risks. Rather, publicly traded companies need to disclose climate-related risks if such risks are material to investors. In Basic, Inc. v. Levinson, 485 U.S. 224 (1988), the Supreme Court explained that a fact is "material" if there is a "substantial likelihood" that a reasonable shareholder would find its omission to alter the total mix of available information significantly. By requiring publicly traded companies to disclose material information, federal securities law enables shareholders to make informed investment decisions.

Subject to this "principles-based" approach to disclosure, a company's management must use its judgment in complying with the "materiality" standard to determine what information must be disclosed. Under the securities laws, investors harmed by materially misleading statements or the omission of material facts can seek remedies through civil litigation.

The SEC's Regulation S-X and Regulation S-K require publicly traded companies to disclose certain information, such as financial statements and business descriptions, in their periodic filings. For example, in the Management's Discussion and Analysis, or MD&A section of SEC-mandated public reports, publicly traded companies must provide a narrative explanation of their financial statements. The MD&A section also requires disclosure of any known trends, events, or uncertainties, which are reasonably likely to have a material effect on the publicly traded company's financial condition or operating performance beyond what its reported financial statements reflect. In the Risk Factors section of SEC-mandated periodic reports, a company must disclose the most significant risk factors that would make an investment in the company speculative or risky, although the rule states that companies should not present risks that could apply generically to any security. Moreover, the SEC's Rule 12b-20 requires additional disclosure of "such further material information, if any, as may be necessary to make the required statements, in light of the circumstances under which they are made, not misleading."

In 2010, the SEC issued a Commission Guidance Regarding Disclosure Related to Climate Change to assist publicly traded companies in satisfying their disclosure obligations with respect to climate change matters. First, the SEC's guidance reviewed certain explicit disclosure requirements concerning environmental matters contained within Regulation S-K. Specifically, under Item 101 of the regulation, issuers must report their material costs of complying with environmental laws. Under Item 103, companies generally must disclose environmental litigation that is material, involves damages in excess of 10% of the company's assets, or involves a government entity.

Next, the SEC provided examples of climate-related risks that a publicly traded company would need to report (including in the business description, MD&A, or risk factors sections), if they are material:

- impacts of enacted or pending environmental legislation, regulations, or international accords that pose a specific risk or a material impact on the company's financial condition;
- indirect consequences of regulation or business trends, such as changes to demand or competition for products or services (e.g., decreased demand for goods that produce significant greenhouse gas emissions) or even relevant reputational effects; and
- significant risks and physical effects of climate-related events such as severe weather.

Although the SEC has not updated its 2010 guidance, it published a "concept release" on Business and Financial Disclosure Required by Regulation S-K in 2016, in which it sought public comment on whether shareholders need sustainability disclosures to understand a publicly traded company's business and financial condition or to participate in shareholder votes. The SEC's August 2019 proposal to amend Regulation S-K does not address climate-related risks.

Voluntary Climate Disclosures. In addition to mandatory disclosures, many publicly traded companies voluntarily disclose climate-related information in corporate social responsibility or sustainability reports. In making these disclosures, numerous companies use disclosure frameworks developed by standard-setting organizations. For example, in 2017, the Financial Stability Board's Task Force on Climate-Related Financial Disclosures (TCFD), an international industry-led group, released its final recommendations report, which outlined a voluntary, standardized climate risk disclosure framework for companies. According to TCFD's June 2019 status update, more than 800 global organizations have shown support for the TCFD framework.

PROPOSALS FOR INCREASED CLIMATE RISK DISCLOSURES

Legislative Proposals. Through hearings and legislative proposals, Congress has been considering possible climate- related macroeconomic impacts and environmental, social, and governance (ESG) disclosures. In the 116th Congress, two bills have been introduced that would require publicly traded companies to disclose specific climate-related information in their annual public reports to the SEC. The Climate Risk Disclosure Act of 2019 (H.R. 3623 and S. 2075), which the House Financial Services Committee ordered to be reported, would require publicly traded companies to evaluate the financial effects of climate- related risks (i.e., both physical and transition risks) and to describe their strategies to mitigate and manage those risks. The bill would require the SEC to adopt specific disclosure rules for certain industries as well as reporting standards for publicly traded companies' greenhouse gas emissions, fossil fuel related assets, and the "social cost of carbon." The bill would also require the SEC to specify a number of line-item disclosures related to greenhouse gas emissions for publicly traded companies developing fossil fuels commercially. Finally, the bill would require publicly traded companies to disclose their financial strategies under different climate scenarios.

Shareholder Activity. Some shareholders have advocated for publicly traded companies to make climate-related disclosures. Under SEC Rule 14a-8, shareholders may recommend votes on proposed company actions at annual shareholder meetings. In recent years, environmental proposals have accounted for a significant share of such shareholder proposals. However, companies often exclude such shareholder proposals from proxy ballots based on the SEC issuing a "no-action" letter recommending that no enforcement action be taken against the company for the exclusion. In addition to making Rule 14a-8 proposals, some investors—through groups such as Climate Action 100+, which represents more than 300 institutional investors—have urged large multinational corporations to make climate-related disclosures using TCFD standards, improve governance around climate change, and take other climate actions.

SHOULD SPECIFIC CLIMATE CHANGE DISCLOSURES BE MANDATORY?

Some proponents of more climate-related disclosures view climate-related risks as having significant, immediate or quantifiable impacts on a publicly traded company's operations and that disclosing such risks aligns with shareholders' interests. Many of these commentators contend that imposing a standardized framework for climate-related disclosures would help publicly traded companies comply with their obligations and enable investors and regulators to compare the risks facing different companies easily. Finally, some observers have argued that enhanced climate-related disclosures could provide regulators with valuable data in monitoring financial system developments at a macroeconomic level.

Critics of increased climate-related disclosure requirements question their usefulness and whether reporting frameworks would capture the complexity of climate-related risks adequately. They also assert that providing such disclosures could increase costs for publicly traded companies—a potentially salient concern in light of the steady decline in the number of publicly traded companies since the 1990s. Some critics also argue that publicly traded companies' disclosures should be limited to information relating to financial performance, viewing additional disclosure as furthering social or political interests rather than providing necessary financial information. Finally, some stakeholders believe that the current disclosure regime (i.e., disclosing material climate-related risks) is sufficiently flexible to account for the climate-related risks publicly traded companies face.

INDEX

A

B

C

D

E

F

G

H

I

J

L

M

R

S

T

U

V

W

Related Nova Publications

A Closer Look at Climate Change

Editor: Reggie Paredes

Series: Climate Change and its Causes, Effects and Prediction

Book Description: This compilation concludes by describing the measures taken by Oman in relation to the environment since the 1970s. The origin of the UNESCO Sultan Qaboos Prize for Environmental Preservation and its establishment in 1989 is also discussed.

Softcover ISBN: 978-1-53614-600-4
Retail Price: $82

Environmental Variations in the North Atlantic in Response to Quaternary Glaciations

Author: Gloria M. Martin-Garcia, Ph.D.

Series: Climate Change and its Causes, Effects and Prediction

Book Description: In this book, the sea surface and deep water conditions in the mid-latitude North Atlantic during the Quaternary are reconstructed based on the analysis of both planktonic and benthic microfaunal assemblages from the SW Iberian margin. The key location of the study site, and the comparison with other sites in the North Atlantic, allow for the drawing of basin-wide conclusions.

eBook ISBN: 978-1-53613-367-7
Retail Price: $82

To see a complete list of Nova publications, please visit our website at www.novapublishers.com